Lecture Notes in Computer Sci<

Edited by G. Goos, J. Hartmanis, and J. van

Springer
Berlin
Heidelberg
New York
Barcelona
Hong Kong
London
Milan
Paris
Tokyo

Willem Jonker (Ed.)

Databases
in Telecommunications II

VLDB 2001 International Workshop
Rome, Italy, September 10, 2001
Proceedings

 Springer

Series Editors

Gerhard Goos, Karlsruhe University, Germany
Juris Hartmanis, Cornell University, NY, USA
Jan van Leeuwen, Utrecht University, The Netherlands

Volume Editor

Willem Jonker
KPN Research
P.O. Box 15000, 9700 CD Groningen, The Netherlands
E-mail: willem.jonker@kpn.com

Cataloging-in-Publication Data applied for

Die Deutsche Bibliothek - CIP-Einheitsaufnahme

Databases in telecommunications II : proceedings / VLDB 2001 International
Workshop, Rome, Italy, September 10, 2001. Willem Jonker (ed.). - Berlin ;
Heidelberg ; New York ; Barcelona ; Hong Kong ; London ; Milan ; Paris ;
Tokyo : Springer, 2001
    (Lecture notes in computer science ; Vol. 2209)
    ISBN 3-540-42623-X

CR Subject Classification (1998): H.2, C.2, K.6, H.3, H.4, J.1

ISSN 0302-9743
ISBN 3-540-42623-X Springer-Verlag Berlin Heidelberg New York

Springer-Verlag Berlin Heidelberg New York
a member of BertelsmannSpringer Science+Business Media GmbH

http://www.springer.de

© Springer-Verlag Berlin Heidelberg 2001
Printed in Germany

Typesetting: Camera-ready by author
Printed on acid-free paper   SPIN: 10840787   06/3142   5 4 3 2 1 0

Preface

Just like the previous workshop at VLDB 1999 in Edinburgh, the purpose of this workshop is to promote telecom data management as one of the core research areas in database research and to establish a strong connection between the telecom and database research communities.

As I wrote in the preface of those proceedings, data management in telecommunications is an interesting area of research given the fact that both service management and service provisioning are very data intensive, and pose extreme requirements on data management technology.

Given the feedback on the previous workshop we decided to keep the same program set-up for this workshop: an invited speaker, a collection of research papers, and a panel discussion. We received 18 good quality papers from which we selected 12 to construct a very interesting program. The program has been divided into four sections.

The first section focuses on CDR data warehouse and data mining technology. Data warehousing and data mining around customer usage data remains an important area of interest for telecommunication operators. The growing competition, especially in the mobile market, means that operators have to put more effort into customer retention and satisfaction.

The second section focuses on performance issues around databases in telecommunication. Since telecommunication databases are characterized by their extreme requirements, for example in terms of volumes of data to be processed or response times, high volume data management and embedded and real-time data management are key aspects of the telecommunication data management problems in today's operational environments.

The third section focuses on database techniques and architectures for the support of data intensive telecommunications services, such as for example broadband services or location services in the context of UMTS. This new generation of services brings new database challenges, such as the modeling and handling of continuous data streams with high quality of service, and the integration of telecommunication and Internet services.

The final section focuses on the embedding of data management technology in the broader perspective of distributed applications and enterprise information management. This is an important topic, since we see a shift from the development of specialized data management solutions by the telecommunication industry towards the application of commercial off-the-shelf technology to the overall information and service architectures.

July 2001   Willem Jonker

Workshop Organizers

Willem Jonker   KPN Research
Peter Apers   University of Twente
Tore Saeter   ClustRa AS
Michael Brodie   Verizon

Program Committee

Heinz Bruggeman	EURESCOM GmbH
Siddhartha Dalal	Telcordia Technologies
Wijnand Derks	Twente University
Dimitrios Georgakopoulos	Telcordia Technologies
Svein-Olaf Hvasshovd	ClustRa AS
Matthias Jarke	Technical University of Aachen
Martin Kersten	CWI, The Netherlands
Steve Laufmann	US West Advanced Technologies
Daniel Lieuwen	Lucent Bell-Labs
Maria de Lorenza	CSELT Telecom Italy
Georgalas Nektarios	British Telecom
Salvador Pérez Crespo	Telefónica
Oddvar Rissnes	Norwegian Telecom
Michael Ronstom	Ericsson
Berni Schiefer	IBM
Martin Skold	Ericsson
Josip Zoric	Norwegian Telecom

Table of Contents

Telecommunications, Databases, and Evolution

Jan A Audestad

Senior adviser Telenor Corporate University
Professor of informatics, Norwegian University of Science and Technology (NTNU)

1. Three Epochs of Telecommunications

From the viewpoint of information storage, the history of telecommunications can be divided into three epochs. The first epoch ended in 1993, take or give one or two years; the second epoch is ending now in 2001. What are then the characteristics of these epochs and why are the awareness of this evolution so important?

Until 1992 telecommunications took place mainly between fixed terminals in the network. The majority of terminals were telephones and the network was designed supporting telephone services and other simple two-person services (telefax, telex, and simple data transmission). The telephone network and early public data networks are static structures consisting of exchanges that are routing and switching the call between source and destination as shown in Fig. 1. All information concerning users,

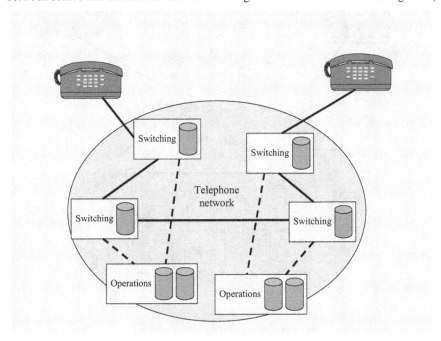

Fig. 1 Telephone network

W. Jonker (Ed.): Databases in Telecommunications II, LNCS 2209, pp. 1-8, 2001.
© Springer-Verlag Berlin Heidelberg 2001

route selection, number analysis and network topology is contained in databases integrated with the switching equipment. The data structures of these databases are simple containing static data structures updated only by remote control from the operations centres. The databases have to fulfil strict real-time objectives (retrieve data within 100 ms) but these are easy to implement. The databases in the operations system have no real-time requirements. They contain usage and charging information, topological information, and customer subscriptions.

In 1992 or thereabout, several events took place that changed the concept of telecommunications. The most important of these events were the implementation of GSM, the development of intelligent networks (IN), and the commercialisation of Internet.

GSM and IN introduced distributed and remote processing of services and calls. The basic network was the same as before. The new element was the remote service control devices (SCP in IN, VLR and HLR in GSM) as shown in Fig. 2. The control devices were required because services such as mobility, freephone and premium rate use nongeographical telephone numbers that cannot be used for routing in the telephone network. These numbers must be converted to numbers with the geographical structure required by the network topology.

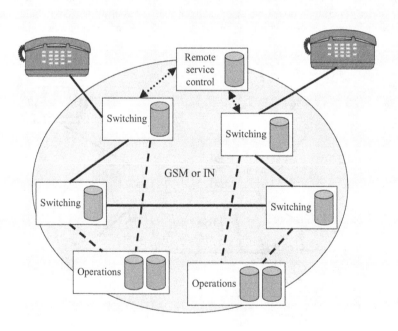

Fig. 2 Remote control in GSM and IN

The remote devices are also used to perform execution of services for the users that were not feasible before. Note also that as one consequence of these developments was that most of supplementary services as we know them today were introduced in the early 1990s. The new requirements introduced by these devices are related to remote real-time management of service processing. Since the devices are inserted as part of the switching process, the time requirement of the remote device is strict. In GSM the databases (VLR and HLR) must respond to service requests within 50 ms where 30 ms is allowed for delay in the transmission system and 20 ms is allocated to database enquiry and processing for individual calls. The databases must be able to handle at least 10,000 independent call events per second. The requirements for IN nodes are equivalent. In GSM, even the access keys to the data in the VLR are random numbers that are changed for every call to and from mobile terminals.

The Internet changed the whole network concept. All service processing in the telephone network is contained in the exchanges. The reason is that the terminals are simple and unintelligent devices. The terminals in the Internet are PCs, servers and other computers. The Internet consists then of a simple network, mainly performing routing of datagrams, while service execution is done in the terminals. The Internet is thus an "inverted" network with intelligence located at the periphery. This way of looking at the network is important in order to understand the evolution of the telecommunications networks now taking place.

The development that is now taking place is to view telecommunications from the viewpoint of the applications themselves and independently of the underlying network structure. This is leading toward new concepts such as peer-to-peer communications, community communication and universal mobility. Below this is referred to as socio-robotic communications.

2. Model of Telecommunications: Networks, Hosts, and Society

Fig. 3 shows a model of the telecommunications network based on IP technology. The system is divided into three layers: the network, the hosts and the social layer representing the users. This model is a consequence of how the interface between IP and the transport protocol (TCP/UDP) is constructed. Let me explain the details of this structure.

The network consists of interconnected router, bridges and ports. This structure offers three things. The basic ability of the network is to transport bits between one access port and one or more other access ports ensuring proper end-to-end quality. IP is used in order to route the data between the ports. Mobility ensures that the terminals may roam between fixed access points (mobile IP) or within a mobile network (WAP, GPRS, UMTS or radio-LANs). This is where the functionality of the network ends. It offers nothing in addition to these basic services.

The hosts (PCs or servers) are interconnected by the transport protocol isolation everything that takes place at the application from the network. The IP – TCP/UDP interface represents a barrier in this structure mush similar to the barrier between the computer hardware and the operating system. The applications that run between the hosts are independent of the network structure except that certain requirements must be fulfilled such as bite rate, residual error probability and real-

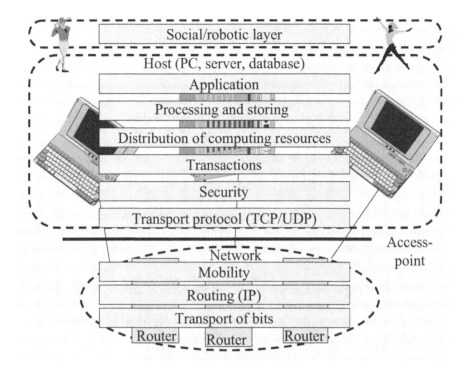

Fig. 3 Model of the new network

time performance. This is similar to the PC where different operating systems may require more or less memory or require minimum clock rate.

At this layer, end-to-end functions of many kinds supporting the application can be included. The figure shows functions such as security, transactions, and distribution of processing and storing of information.

The social or robotic[1] layer makes use of the end-to-end functionality of the layer below. The World Wide Web exists at the socio-robotic layer; and so does delivery of comfort to homes by remote reading of thermostats, detection of presence or absence of the residents, and switching ovens on and off. The layer below the socio-robotic layer supports protocols and processes such as http, multimedia management, and automatic payment of services required by the services at the socio-robotic layer. Again, we observe that there exists a barrier between the socio-robotic layer and the layer below. This is again equivalent to the barrier between the operating system of a computer and the application programs that can be run on that operating system.

Functionality of the three layers must then develop independently of one another. It will be a mistake to integrate this functionality in a vertical structure. This

---

[1] Social for communications between people; robotic for communications between machines.

is obvious in computer science since we do not develop a separate computer for each application program we design. However, it is not obvious in telecommunications since traditionally this is the way in which telecommunications systems was constructed until the 1990s – separate networks for telephony, telex, mobile services and data transmission.

3. Distributed Architecture: Mobility, Webs, and Communities

Let us now look at some of the impacts of this development. These are impacts that all actors in the industry must be aware of: telecom operators, computer manufacturers, application designers and information providers.

GSM offers one particular type of mobility, that is, telephone services while user is moving. The idea of UMTS is to extend this mobility to also include data transmission. One criticism of UMTS though is that the design is nevertheless based on telephony and not on data transmission. Different radio-LAN designs offer similar capabilities based on IP technology.

PCs can also be moved between different access points, however, with severe restrictions due to security (e.g., firewalls). All this is then concerned with mobility of the terminal (and the user together with it). The data management problem in all these cases are similar to that of GSM and mobile IP. The mobility is offered by the network layer and it is well understood how it can be implemented.

The next step is to allow users to move between different terminals retaining her or his services. This type of mobility belongs to the layer of hosts. It must therefore be implemented at some layer above TCP. Platforms supporting this type of mobility are still not generally available. Technologies that may be used for personal mobility are intelligent agents, mobile agents and CORBA. The data management problem is then to maintain relationships between the user location and the user profile identifying how the user services are to be executed.

Session and application mobility represents an even more difficult problem. This implies that the user may leave a session at one terminal and later pick it up at another terminal without going through normal logout and login procedures. One difficulty is that this type of mobility really belongs to the socio-robotic layer. Therefore, it should not be designed in such a way that it presupposes or requires any particular design of the network layer or the host layer.

In a truly distributed socio-robotic architecture, all three types of mobility are required.

The World Wide Web is independent of the Internet. Usually we believe that it is the other way round. In order to understand this, we must observe that TCP (or UDP) offers an unstructured stream interface to the processes using it. The services TCP offers to the user process are concerned with quality-of-service and integrity of data. Therefore, the web can be designed on top of any end-to-end protocol (or network protocol) offering a stream interface. The evolution of web systems is thus not linked to the evolution of Internet.

The web is an interconnected structure. The interconnection address is the hypertext and two web pages are interconnected if there is a hypertext pointing from one page to the other. Search engines extends this interconnection structure to comprise other address elements (single words or combination of words in the web files).

The web consists of mainly unordered structures of passive files. The web is not at present used for executing complex dynamic programs. However, the web is a mostly unordered structure.

Communities can be defined as ordered structures based on addressing mechanisms similar to that of the web. A community consists of persons with some common interest. The common interest may be related to their work, family, friends or hobby. The participants may share files, address books, appointment books, diaries and executable software. The concept requires that the users must be able to access the community at any fixed or mobile access point via different types of terminals and interface. The community management system must take care of the capabilities the current interface may provide so that the service options of the user is dynamically updated so that, for example, video is not sent via narrowband GSM accesses. The communities are advanced systems offering all levels of mobility described above in addition to security and protection of integrity. Membership of the community and access rights to information may change dynamically. The community can consist of members worldwide. However, the biggest challenge is not communications within the group but the management of the complex and dynamic data structures associated with the community itself and its individual members. What makes the challenge even bigger is that the community must not violate restrictions associated with individual members, for example, disclosure of information owned by other parties such as employees or violation of intellectual property rights.

The exploitation of community communication has just started but is expected to become one of the major driving forces in telecommunications soon.

4. Communities of Machines: Remote Sensing and Control

The communities described above consist of people. Similar ideas can be employed to communities of machines. A community of machines can consist of cars, the manufacturer of the cars, road authorities and repair shops licensed for selling and repairing the cars. The communications within the community can be the car sending fault alarm to the repair shop, the manufacturer performing remote testing of the car, the car requesting roadmaps and driving directions from the road authorities, and the manufacturer updating the software of the car's computers. Patients carrying medical sensors in or on their body, remotely monitored by hospital computers, can also be viewed as a community of machines. Intelligent buildings together with electrical power plants, alarm companies and waterworks may also form a community with several interests combined in one system.

The key point here is the availability of cheap and small sensors. Sensors such as accelerometers, gyros, spectrometers, hygrometers, thermometers and light-detectors (some of them even include CPUs and communications devices) can be built in volumes as small as one cubic millimetre. These devices are called microelectromechanical systems (MEMS) and are of particular interest in energy management systems, alarm systems, automotive industry and medical industry.

The application of MEMS in the auto industry is evolving fast. The cars are also equipped with communication devices – at present GSM terminals but in the near future also IP-zone communication. It has even been proposed that cars should be equipped with packet radio stations that communicate between themselves and with fixed base stations along the road. This will produce a moving radio Internet offering stochastic interconnectivity. Managing such a network is indeed a challenge!

Remote sensing and control is assumed to become one of the major users of telecommunications in the near future. Today many of us accesses the telecommunications network via several subscriptions: fixed telephone service at home, internet access at home, GSM terminal, internet access at work and the fixed terminal at work. Application of remote sensing and control will increase this number to perhaps ten or more accesses per person: our home, our car, our hart and so on. In addition, all infrastructure contains already or will contain soon devices for remote sensing and control.

All this connectivity can easily be supported by the IP network technology. However, the real challenge is to design and implement all functions that are required at the layers above IP: between the hosts and between members of socio-robotic communities.

An additional challenge is related to the traffic characteristics of the new applications. Speech is well approximated with a birth-death Poisson distribution function. The traffic distribution in Internet shows already sign of becoming fractal and asymmetric (much more terminating than generated traffic). Therefore, traffic dimensioning of future network will require new methods.

The ultimate miniaturisation of sensors and machines is nanotechnology. Micro-miniaturisation is approached from two directions: standard graphical reproduction as in VLSI, and building devices atoms by atoms. The purpose of both technologies is to make small and versatile components that can be used in places and for applications not possible with current technology. The limitation of the first approach is the granularity that is possible even with X-ray imaging. The most important limitation of the second approach is to find appropriate building methods applicable on the atomic level. Several techniques have been exploited: self-assembling molecular structures, positioning single atoms by using scanning microscope technology, and constructing molecular conveyor belts – inspired by cell biology – for assembling molecules in a predetermined manner.

Nanotechnologies are still far into the future. However, breakthrough may come fast and unexpectedly. Therefore, nanotechnology cannot be ruled out as science fiction. Gene manipulations may be one way in which bacterial cells are programmed to produce nano-robots or other sophisticated devices. Reproduction of the way cells are producing proteins is another feasible approach.

5. Advances in Computing

Finally, I will mention some developments in computing that may have impact on telecommunications. Moore's law predicts that the computing power of PCs doubles approximately once a year. The increase in the demand for telecommunications seems now to be well correlated with the increase in computing power of PCs, though at a much slower rate. On the other hand, is this increase caused by the increase in computing power or is it caused by the decrease in both size and price of computing devices – making them available for more people? However, as telecommunications operators we cannot just ignore this development.

Moore's law predicts that by 2020 the size of semiconductor storage devices will have the size of a single atom. This limit is of course not reachable with standard ways of producing micro-miniature devices. Long before this limit is reached, quantum mechanical effects have put an end to the way in which macroscopic bodies behave. What remains then is to construct the devices on the quantum mechanical foundation.

The basic principle behind quantum computation is that a quantum system may consist in several superimposed states at the same time. It can be shown that this makes it possible to construct – in theory at least – Boolean circuits resembling those in ordinary computers. Having constructed such circuits, they can be combined to perform computations in a way resembling that of other computers. The concept can be brought further by observing that an array of n atoms can be employed to construct a superimposition of 2^n states. Several algorithms have been designed to perform certain types of computations such as factoring large numbers and searching for a given item in a list. These are problems that are "untractable" for ordinary computers, meaning that the time taken to solve the problem increases sharply with the number of digits in the number to be factorised, or the number of elements in the list. Such problems can be solved immediately by the quantum computer.

Still no one has constructed a quantum computer containing more than two Boolean gates. The difficulty is related to preserving quantum states long enough to make computations on them, to pass information from one quantum system to another, and to design input/output devices that do not destroy the quantum state. Bearing in mind the speed by which experimental technologies are developing in physics, it is no reason to believe that these problems cannot be solved.

Other powerful computing devices are based on biology. These include neural nets (artificial brains), DNA computing manipulating with sequences of amino acids, and membrane computing utilising the enormous potential of cell membranes to recognise and respond to molecules attached to their surface.

The connectivity of the World Wide Web – or rather the one that is developing for community communication – resembles a neural net with powerful standard computers as nodes rather then neurones. The web is much smaller than the human brain but it is indeed much larger than the neural system of an insect. The insect can use these resources for complex processes such as to fly, walk, hunt, hide and reproduce. Research is no going on concerning how the web can be exploited to make computations of a similar or larger magnitude of complexity possible.

Data Warehouse Population Platform

Jovanka Adzic, Valter Fiore, Stefano Spelta

Telecom Italia Lab
Via G. Reiss Romoli, 274
10148 - Torino, Italy

{adzic, fiorev, spelta}@telecomialialab.com

Abstract. The input data for a data warehouse, coming from operational systems, are not immediately ready for loading into data warehouse. It may need cleaning and integration with other data. Moreover, the input data must be transformed and translated into a format more suitable for analytical purposes. This paper presents a generalised platform for population of data warehouses named *Data Warehouse Population Platform* (DWPP), a set of modules whose aim is to resolve typical aspects arising during the transformation and loading vast amount of data into data warehouse. Using DWPP modules application developers have to add-on only domain specific transformation logic and integrate all in a new Population System. In Telecom Italia Lab we have realised several Population Systems based on DWPP. The Case Study section describes one of them: Population System for Mobile Network Traffic Data Warehouse.

1   Introduction

Business and telecommunication systems have to collect vast amounts of detailed data such as Call Detail Records (CDRs). These CDRs are used, first of all, for billing purposes and secondly, they are often stored in decision support databases, very large databases, in order to answer different needs. There are a few areas that need fast and reliable access to very large databases. For example Customer Relationship Management (CRM) and Business Intelligence have to deal with large volumes of data containing a lot of information about a company's business and behavioural patterns of the company's customers. The challenge is how to deal with this vast amount of data, how to store and analyse it, in order to discover valuable information about business and customers.

The input data from an operational system, very often, are not immediately ready for loading into data warehouse. Firstly, the input data may need cleaning and integration with other data. Secondly, the input data must be transformed and translated into a format more suitable for analytical purposes. The pre-processing of detailed data (CDRs) is necessary before loading into data warehouse. Techniques such as Extracting, Transforming and Loading (ETL Tools) support transforming and loading data into a data warehouse. Techniques such as On Line Analytical Processing (OLAP

W. Jonker (Ed.): Databases in Telecommunications II, LNCS 2209, pp. 9-18, 2001.

Tools) and Data Mining (DM Tools) support analysis of data and finding of hidden patterns in data stored in data warehouse.

The contribution of this paper is to bring our past experience in developing and deploying different data warehouse Population Systems. We had evaluated some ETL tools, but we could not find one corresponding to performance requirements in transforming and loading very large amount of data (CDR Transformation & Loading). This paper presents a generalised platform for population of data warehouses named *Data Warehouse Population Platform (DWPP)*, a set of modules whose aim is to resolve typical aspects of transformation and loading vast amount of data into data warehouse. DWPP does not concentrate on extraction of data from source systems, but assume to have input data already exported or produced in flat file format. *DWPP* is tuned for data warehouses on Oracle DBMS and it is independent from the domain of data (CDRs, Customer Data, Transaction Data,…).

Using DWPP "already done" modules application developers can efficiently resolve some of typical questions in data warehouse population (Synchronization, Shared Memory Management) so they can concentrate on core question - transformation of input data - and produce a Population System for a very large data warehouse in short time. Using DWPP modules application developers have to add-on only domain specific transformation logic and integrate all in a new Population System.

Outline

The rest of the paper is organized as follows: Section 2 describes DWPP principal features, such as dispatching input data, synchronizing application transformation units, logging activities, loading output data into Oracle database tables, etc. Section 3 presents a scenario of a Data Warehouse Population System, based on DWPP, for Mobile Network Traffic Data Warehouse that we have deployed for an important mobile operator. Section 4 describes some new features coming in the next release of Data Warehouse Population Platform.

2   Data Warehouse Population Platform Features

The *Data Warehouse Population Platform* (DWPP) is a set of C software modules on Unix S.O, that include features such as:
- Identification, collection and distribution of files containing input data for the transformation units;
- Synchronization of data transformation units;
- Fast lookup functions for surrogate key valorisation;
- Bulk data loading into database tables.

The DWPP Modules are shown in Figure 1. The upper levels are extensions of the lower levels that expose an API and allow the platform to be independent from application logic. The first level modules wrap Operating System and Oracle DBMS, the second level modules are extensions that offer particular features for data warehouse population.

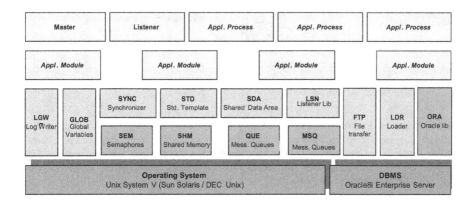

Fig. 1. Data Warehouse Population Platform Modules: *SEM, SHM, QUE* and *MSQ* wrap Operating System; *ORA* warps OCI (Oracle Call Interface); *LGW* is used for logging operations; *GLOB* is used for managing global variables; *FTP* wrap File Transfer Interface; *LSN* is used for input data dispatching; *SDA* is used for managing shared in-memory data area; *SYNC* is used for synchronizing transformation units; *STD* is standard template for different kind of transformation units and *LDR* is used for transforming and loading data

The typical architecture of a populating system based on DWPP is shown in the following Figure 2.

The **Master** process allocates shared resources used by all process units and takes care of their synchronization.

The **Listener** process collects data flows, File Group, coming from source systems and dispatches them towards appropriate data transformation units.

The **LogWriter** process writes log messages on the system log files about all activities done during population phase.

The **Shared Data Area** (SDA) is a main memory segment used by the application transformation units. The SDA is divided into many sub-segments whose access can be restricted to some units or be made public. The SDA offers the basic functions of an In-Memory Data Base allowing fast search in lookup tables that works as a cache for database tables. This feature is heavily used during surrogate key transformation, done by transformation units.

The **Appl. Process** can contain different application specific transformation units which transform the input data according to a specific application logic.

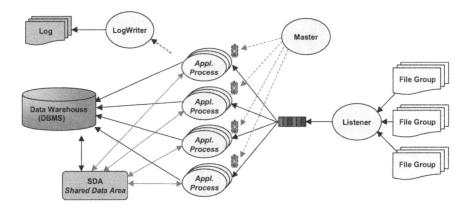

Fig. 2. Standard Architecture of a Populating System based on DWPP: *Master* is a coordinator of all system activities; *Listener* is dispatcher of input data to transformation units; *Log-Writer* writes to log about all activities done during population; *Appl. Process* is particular transformation unit (process or thread); *SDA* is a main memory segment used by transformation units; *File Group* contains input data

LSN Module – Input Data Dispatching

The Listener process takes care of collecting data-flows coming from source systems and dispatching them to data transformation units. A data-flow can be made of many different files and there can be different criteria to determine the flow termination. DWPP identifies each flow as a File Group assigned to a particular ProcessUnit. Each File Group must be characterized by:

- The incoming directory where Listener looks for files;
- The Process Unit to which File Group files must be sent;
- The file name pattern;
- The waiting timeout;
- Pre-processing operation that Listener must compute (uncompress, split, header and footer skip);
- The number of threads dedicated for pre-processing operation
- The flow termination criteria (maximum number of files, timeout, custom function).

The Listener process waits for files belonging to the defined File Groups, until flow termination is reached. As soon as Listener process receives the file the assigned ProcessUnit is notified through a message on a shared message queue. All instances of ProcessUnit cyclically check the presence of messages on the queue. On termination of a File Group the Listener sends on the queue one termination message for each instance of the ProcessUnit. As an instance of ProcessUnit reads a termination message it assumes data-flow termination and exit with success.

Input data files belonging to a File Group can reach different states depending on the transformation phase they are in. For each state, there is a corresponding subdirectory in the main incoming directory. These states are named:

- incoming: the input data file as transferred from the source system to the incoming directory where the Listener waits for it;
- queued: the input data files, eventually pre-processed by Listener (decompressed and/or split), ready to be processed by the appropriate ProcessUnit; Listener process informs that Process Unit by sending a message via shared message queue.

The Listener process must be implemented "ad hoc" to reflect data flow characteristics of a DWPP based population system. LSN Module offer API interface for defining File Group.

SDA Module – Shared Data Area

The *Shared Data Area (SDA)* is a main memory segment used by the application and system processes to maintain useful data. The DWPP module named SHM exposes an API to manage shared memory accesses. Allocation and de-allocation of SDA is done by Master process and all other processes just have to gain access to it. Application processes can allocate memory sub-segments inside SDA. These sub-segments can be private (only the creator process can access them) or public. Sub-segments allocation, managed by the SHM Module API, allow application processes to obtain memory segments without accessing to OS level.

The Shared Data Area always contains two private sub-segments: one used only by Master process for synchronization purposes, and one used only by Listener process for input data dispatching purposes.

One of the very useful features in a data warehouse population is the possibility to transform keys used by source systems into (surrogate) keys used inside the data warehouse. This transformation involves a fast search on a lookup table. The time consumed by this operation is critical in high volume population systems. DWPP allows the managing of lookup tables in a public segment of SDA, called *Rapid Search Area (RSA)*. This segment is allocated by an "ad hoc" process, which also takes care of its initialization through a set of queries, executed on the database. This process, often called *InitRSA*, must complete the initialization of the lookup tables area before the starting of any application ProcessUnits.

Each record of the lookup table is a structure containing all fields needed by the data transformation functions, so there is no need to access DBMS during transformation. All data necessary for transformation are cached in SDA from data base.

A lookup table can be organized for binary search (ordered and only for read access) or for hash search (not ordered and for read/write access). It is possible to define lookup table structures that combine different organization, in order to avoid locking the whole lookup table for every access. This lookup table is made of a Primary table and an Overflow table. The search key to allow a fast binary search sorts records in the Primary table. Records in this area are read-only and loaded during the initialization phase. Records in Overflow table are not sorted and are retrieved through a sequential search. New records can be added or old record can be updated in overflow table. An exclusive lock protects only the Overflow table access.

To define a lookup table in RSA it is necessary to specify the following information:

- Record structure of the lookup table;
- SQL query to load Primary table;
- Table indexing type (BINARY or HASH);
- Access Type;
- Compare function or hash function to establish equality between the search key and the record key;
- Insertion function to insert new records in the Overflow table;
- Number of records estimated for insert into Overflow table

The search algorithm on a lookup table made of Primary and Overflow table follows these three steps:

- Primary table search through binary search algorithm. If the search key is found the record is returned, otherwise the next step is executed;
- Overflow table search through the sequential search algorithm after obtaining exclusive access. If the search key is found the record is returned, otherwise the next step is executed;
- Overflow table insertion (if the process requests it and free space is available). The record is returned.

The search algorithm on a lookup table made of Primary table only follows the next single step:

- Primary table search through binary or hash search algorithm. If the search key is found the record is returned, otherwise the transformation have to be completed in a different way (default values or the input record will be rejected and re-elaborated again in the next execution of population).

The DWPP module named RSA (Rapid Search Area) exposes an API to manage lookup tables: definition, initialization, data retrieval and data update.

SYNC Module – Units Synchronization

The population system based on DWPP is made of elementary pieces, called ProcessUnit. The ProcessUnit can be Unix-process or thread implemented and multi-instantiated to increase parallelism. All instances of a ProcessUnit are activated and work contemporarily. The Master process is responsible for activation, initialization and synchronization of all ProcessUnits and tasks during the population phase.

Synchronization is defined at ProcessUnits level. If a ProcessUnits P_i depends on ProcessUnits P_k ($P_k \rightarrow P_i$) that means that all instances of P_i can begin working when all instances of P_k are terminated with a particular termination state. With the definition of dependency rules among ProcessUnits, it is possible to specify also the expected termination states of the ProcessUnits another ProcessUnits depends on. The SYNC module allows the definition of complex synchronization rules (AND / OR) that determine dependency matrix.

Each instance of a ProcessUnit can reach different states:
- QUIET: state during system startup prior to allocation and initialization of data structures;
- READY: state after the initialization phase; waiting for the termination of the ProcessUnits it depends on;
- STARTABLE: state when all instances of ProcessUnits it depends on are terminated; it has the permission to start;
- WORKING: state during the data transformation phase;
- SUCCESS: termination state of a successfully terminated instance;
- WARNING: termination state of an successfully terminated instance but with some warnings;
- TIMEOUT: termination state of an unsuccessfully terminated instance due to insufficient time to finish transformation;
- FAILURE: termination state of an unsuccessfully terminated instance due to a failure that has interrupted transformation phase.

The Master process, containing the definitions of ProcessUnits synchronization rules, must be implemented "ad hoc" to reflect specific populating systems dependency rules. SYNC Module offer API interface for defining dependency rules.

LDR Module – Transforming & Loading Data

All DWPP application process use the loading module (LDR) to transform and load data into the Data Warehouse tables. The application processes contains transformation functions (ProcessUnits) plugged into the LDR module (Figure 3).

The LDR module is one of the most important and most performing module of DWPP. It has an internal multithreaded structure based on:
- Input Buffer Pool – private memory segments used by Readers (from Readers Thread Pool) to write input data;
- Output Buffer Pool – private memory segments used by Elaborators to write transformed data ready to be loaded by Writers (from Writers Thread Pool);
- Reader Thread Pool - threads that open the input data files, once that the appropriate message has been detected from the Listener, and spread that data over the Input Buffer Pool;
- Elaborator Thread Pool – threads that effectively transform the input data and move it from Input to Output Buffer Pool;
- Writer Thread Pool – threads that move transformed data from Output Buffer Pool to database tables or flat files.

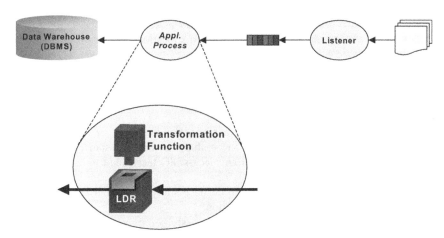

Fig. 3. Architecture of an Application Process: *Transformation Function* has to be defined and plugged into *LDR* Module in order to define one transformation unit

The transformation function that has to be plugged into LDR Module is the core element of Elaborator's code (DWPP offers lot of pre-defined functions/macros useful for transformation logic).

The LDR Module can contain different kind of transformation function, and so different kind of Readers, Elaborators and Writers. The sharing of the Input and Output Buffer Pool between different kind of transformation function is managed by LDR Module concepts: Input Channel – logical flow of data that comes in the ProcessUnit (Elaborator) and Output Channel – logical flow of data that comes out of the ProcessUnit (Elaborator). One ProcessUnit can have one Input Channel and multiple Output Channels.

LDR Module offer API interface for defining Buffer Pools, Channels and ProcessUnits.

3   Case Study

The population system for Mobile Network Traffic Data Warehouse has been developed using the Data Warehouse Population Platform (DWPP) and deployed for one of the biggest mobile network operator. This Data Warehouse is aimed to store cellular network traffic data (CDRs), billing and other customer data in order to enable different kind of analysis, principally off-line fraudulent behavior analysis.

The Data Warehouse Population System elaborates, day by day, the entire mobile network traffic, customer data and data coming from on-line Fraud Detection and Management System. The Data Warehouse population system is able to transform and load about 80 million CDRs per hour.

The Data Warehouse, based on Oracle8i, is made up of different federated logical Data Marts. There are two different sub-systems related to different customers segments: subscribers and pre-paid cards. The entire Data Warehouse is about 1 TB, 300 GB related to subscribers and 700 GB related to pre-paid cards.The database server is Oracle8i Enterprise Server on a Digital Unix Alpha Server 4100.

The Data Warehouse are organized in star schemas with dimensional and fact tables. The principal dimension tables are: Customer, Location, Direction, CDR Type, Time. The fact tables contains traffic detail data. All fact tables are partitioned by value and hash sub-partitioned due to:

- "managing history" needs - every day you have to add new partition and drop the oldest; you have to export for backup needs only the new partition;
- performance needs - Parallel Query Servers work on different sub-partitions.

The main fact table, containing traffic detail data of all customer, has about 2 Billion of records. The Customer dimension table are also hash partitioned due to volumes of data (about 22 Million of Customers).

Data stored in the Data Warehouse are accessed by different kind of analytical tools in order to monitor traffic trends (risk directions) and discover patterns that describe or predict fraudulent behaviors. The OLAP tool is Oracle Discoverer; the Data Mining tool is SAS Enterprise Miner. Oracle Developer has been used to develop the "ad-hoc query" client application used for browsing detail data about customer.

4   Future Works

The *Data Warehouse Population Platform* (DWPP) has been used in deploying different data warehouse population systems on Unix S.O.

In the next phase, we plan to add new features to the DWPP. One of that new features will be a *Data Warehouse Administration (DWA)* Module providing useful API interface for "managing history" in a data warehouse. At the moment, the managing history logic is at the application level.

In data warehouse environment "history" and "time" are unavoidable factors to consider at the data warehouse design time. Incremental loading into a target table which cumulates historical data (usually fact table) scheduled on a regular basis are typical situation in data warehouse environment.

In our experience we realized that data partitioning can be very useful in managing history from different point of view:

- expected high performance that a data warehouse could bring to business application (OLAP & Data Mining) that often need to query a large amount of historical data along a time dimension - the data partitioning can effectively help the queries cut down the amount of data that have to be processed (automatically done by Oracle SQL Optimiser – partition pruning);
- removal of dormant data aged out over time – the data partitioning is enabling technology because you can drop "the oldest partition";

- backup - the data partitioning eventually become a lifesaver for many data ware-house; export only "just loaded fresh data" instead of using classical backup utilities;
- high performance during the loading of data warehouse – the data partitioning can help with parallel direct-path insert into a temporary table; whenever an error occurred during loading, the temporary table can be easily dropped and error logged; except for the final partition exchange step the target table is always available for any database operations.

How to partition data is therefore a key data warehouse design issue. The DWA Module will provide APIs for defining different type of partitioned tables, related indexes, time period that you have to keep the data in data warehouse and it will be based on some strict naming convention.

In the next phase we will also investigate issues that arise assuming input data for data warehouse in XML. In a very short time, most Web data may be stored and inter-changed in XML.

5   References

1. Ralph Kimball, et. al., "Data Warehouse Life Cycle Toolkit", John Willey & Sons, Inc., 1998 ISBN: 0-471-25547-5
2. Dr. Yu Gong: "Partition Exchange Loading: A new Performance feature in Oracle Ware-house Builder, Release 3i", An Oracle Technical White Paper, May 2001:

Experimenting NT Cluster Technology for Massive CDR Processing

J.E.P. Wijnands, S.J. Dijkstra, W.L.A. Derks, W. Jonker,

KPN Research, PO Box 15000, 9700 CD Groningen, The Netherlands
{j.e.p.wijnands, s.j.dijkstra,
willem.jonker}@kpn.com
derks@cs.utwente.nl

Abstract. This paper describes the final results of an assessment of the scalability of an NT Cluster architecture for very large Telecommunications Call Detail Record (CDR) databases[1]. The IBM DB2 DBMS was used on top of four standard Windows NT multi-processor computers to investigate the data warehousing capabilities of such an off-the-shelf solution. First a description of the experiment case and experiment set up is given. Next the results of the experiments are described. The paper ends with a conclusion on the manageability, robustness and scalability of the NT Cluster architecture for very large CDR stores.

1   Introduction

The generation of so called Call Detail Records (CDRs) by modern telecommunication switches has opened the way to a whole area of new applications. Most importantly CDRs are the basis for almost all operational billing systems today. And recently, with the advance of database technology for very large databases, new applications based on the analysis of huge amounts of CDRs (tens of millions a day) have come within reach. Examples of such applications are traffic management, fraud detection, and campaign management.

Given the large amounts of data (terabytes of CDRs are normal) there is a need for very large and high performance databases. One way to accommodate these requirements is to use specialised high performance database servers such as for example IBM mainframes, Compaq (TANDEM) or TERADATA machines. However, telecommunication companies are hesitating to make large investments in technologies supporting these new applications. What they are looking for is cost-effective and scaleable technology for realising very large databases in an incremental way. Given that there is a general trend of commodity hardware and software becoming more and more powerful with respect to both processing power and storage capacity, the question arises whether this technology is mature enough to support the specific CDR analysis applications of telecommunication companies.

---

[1] The work described here was partly carried out in the context of EURESCOM project P817 "Database Technologies for Large Scale Databases in Telecommunication" (www.eurescom.de).

W. Jonker (Ed.): Databases in Telecommunications II, LNCS 2209, pp. 19-36, 2001.

To answer this question we performed a state-of-the-art study on scaleable architectures for CDR stores (see [1]). There we concluded with two promising solutions: NUMA and NT clusters. Following the state-of-the-art study, we experimented with both technologies to assess the scalability, manageability and robustness of both solutions. After a market survey, IBM was chosen as preferred supplier for both technologies (see [3] and [4]). This paper describes the results of experiments, carried out in the first two quarters of 2000, with a CDR store using DB2 UDB EEE on IBM Netfinity 7000 machinery. For results of the experiments with Oracle 8.0 on NUMA machinery see [2].

2   Case Description

As case for performing the experiments, we took the Marketing Customer Data Base (MCDB) of KPN Mobile as an example. This data warehouse contains the CDRs and customer data of all customers of KPN Mobile. The system only has 1 to 5 concurrent users with very heavy marketing and management queries. The one-user results are described in this paper.

2.1   Cluster Architecture

The cluster architecture is a *Shared Nothing Architecture* consisting of multiple *Symmetric Multi Processing* (SMP) nodes, all having their own disks, memory, CPU's and image of the operating system. A fast TCP/IP LAN connects these nodes with each other.

2.2   Database Schema

The NT Cluster system was investigated primarily for its data warehouse potential. The data we used included a large CDR table and detailed customer data. The customer data included name, address, and detailed product, order and delivery information. In this case, a large CDR table means that it accounts for 90% of the total data volume. The data schema is shown in Fig. 1.

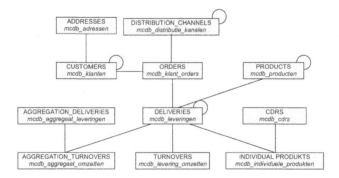

Fig. 1. Simplified data schema of the used application

2.3   Application

The application running on the NT Cluster is executing various queries against different system configurations using an SQL interface. Eleven different real-life marketing queries were executed containing both join and group-by operations. Note that we investigated read-only queries and focussed on intra-query parallelism. The business questions and the SQL queries of the most relevant queries (i.e. queries involving the huge CDR table) are described below and will be further discussed in this paper.

The GBA (group-by on A-number (also called ServedMobileNumber)) query answers the business question: *"What is the total call duration per A-number (=the calling party) for all calls in the so-called peak-hours (between 07:00 a.m. and 08:00 p.m.)?"* The SQL statement we build for this question is:

```
SELECT SERVMOBNR, SUM(decimal(CALLDUR))
FROM CDR
WHERE hour(STRTCHRGDATE) between 7 and 20
GROUP BY SERVMOBNR
```

The GBB (group-by on B-number (also called OtherPartyNumber)) query answers the business question: *"What is the total call duration per B-number (=the called party) for all calls in the peak-hours?"*
The SQL statement we build for this question is:

```
SELECT OTHERPRTYNR, SUM(decimal(CALLDUR))
FROM CDR
WHERE hour(STRTCHRGDATE) between 7 and 20
GROUP BY OTHERPRTYNR
```

The MJ (multiway join) query answers the business question: *"What are the last names and call duration's for all customers?"* This query is a bit more complex because the CDR has to be joined with several customers related tables to get the last names (Achternaam):

```
SELECT ADS.ACHTERNAAM, CALLDUR
FROM LEVERINGEN LVG, KLANTEN KLT, ADRESSEN ADS, CDR,
INDIVIDUEL_PRODUKT IPT
WHERE CDR.SERVMOBNR=IPT.KENMERK AND
IPT.LVG_ID=LVG.ID AND
date(CDR.STRTCHRGDATE) BETWEEN LVG.DATUM_BEGIN AND
LVG.DATUM_EINDE
AND LVG.KLT_ID=KLT.ID AND
ADS.KLT_ID=KLT.ID AND
ADS.TAS_CODE='BEZOE'
```

Finally, there is the MQ (Marketing) Query combining join and group-by operations on a small customer related table only.

3   Configuration

3.1   Approach

The primary aim of the experiments is to reveal the actual scalability behaviour of an NT cluster architecture *in practice*; starting with a small data set and accompanying hardware, can the hardware keep pace as the data sets grows. Therefore we configured the system initially with a small database (100 Gbyte) with limited system power (1 node) and extended the database and system eventually towards 400 Gbyte with 4 nodes. To explain the measured system behaviour, additional experiments were performed to reveal the impact of specific system resources on the scalability behaviour (i.e. CPU, memory, interconnect). In total we performed about 250 experiments. Beside this quantitative analysis of Scalability, we also did a qualitative analysis of Manageability and Robustness.

3.2   Cluster Configuration

The cluster we investigated consisted of (up to) four IBM Netfinity 7000/M10 systems interconnected by a Netfinity SP switch (see Fig. 2).

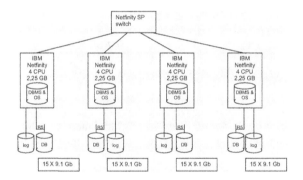

Fig. 2. Four node cluster hardware configuration

The systems each run their own image of Windows NT4.0 (service pack 4). Each machine has four Intel Xeon 450 Mhz CPUs, 2 GB memory, two 4.5 GB disks and sixteen 9.1 GB disks. This results in an overall capacity of 16 CPUs, 8 GB memory and 618,4 GB of raw disk space. The 4.5 Gb disks are configured RAID-1 and contain the OS and DBMS software. The other disks contain data and are configured RAID-5 with 5 physical disks per logical disks. So there are three logical disks per node (see Fig. 2). On each node, the remaining disk is used for logging and load purposes.

3.3   Database Configuration

IBM's DB2 UDB EEE version 6.1 (called "DB2" in the sequel) was used as DBMS. From a hardware and operating system point of view, the cluster consists of 4 separate machines (see Fig. 2) but from a database point of view, DB2 provides a single image of the database. So DB2 contains software to cluster the components and make the total system behave as one.

In a shared-nothing environment, where each node manages its own subset of the data, the distribution (also called *fragmentation* or *partitioning*) of data over the available nodes in the cluster is very important. This distribution of data can be influenced/determined by the database designer (unlike for example the query optimisation strategy). Fragmentation strategies can be divided into two categories: horizontal and vertical fragmentation. *Horizontal* fragmentation distributes complete records of a table to different nodes. Thus, the table is cut into pieces by horizontal slicing. *Vertical* fragmentation cuts vertically through a table, separating the columns of a table. Note that the fragmentation of a data set can also be based on a mix of both strategies. This is called *mixed* fragmentation. When data is fragmented, the allocation of the fragments to the nodes may impact the performance of the query evaluation. In particular, the join of two fragments on different nodes will be more expensive than the join of two fragments on the same node. The former situation is called non-colocation and the latter is called colocation. As co-location is a relevant parameter for shared-nothing architectures, we have included it in our experiments.

Horizontal fragmentation is supported by DB2: a table can be hashed over (some or all of) the nodes in the cluster based on a so called *Partition Key*. The partitioning key is a set of one or more fields of a table. Vertical fragmentation is not supported by DB2. Although vertical fragmentation could be simulated by splitting tables manually, we decided to only consider horizontal fragmentation in our experiments. The large CDR table was horizontally fragmented over all available nodes in the cluster using the ServedMobileNumber (the phone number) as partitioning key. Also the IPT table, used to directly join with the CDR table, was horizontally partitioned on the same attribute. The rest of the tables is relatively small, and we decided not to fragment them. Instead we replicated them over all available nodes. In DB2 this is easy, because this replication is supported transparently to the user.

In addition to the data, also the index and temporary storage is spread over all nodes. Hence, all nodes in the cluster are identically configured. Within each node, the DATA, INDEX and TEMP tablespaces are divided over all disks as depicted in Fig. 3. The index is fragmented according to the fragmentation of the data. Also the TEMP tablespace is available on each node, so maximal local processing is possible.

Fig. 3. Division of data over the disks on a node

3.4   Experiment Configurations

For the experiments, the hardware and database settings were varied. With respect to hardware, the number of nodes and the configuration of each node were varied. Each node is equipped with four CPU's, a memory unit and 15 RAID-5 configured disks. For our experiments, we varied the amount of memory per node, the number of CPUs per node, and the number of nodes in the cluster.

Besides these hardware parameters we also changed the database size. To describe the combinations of hardware and database sizes, the following notation is used: P<xxxx>DB<y>

xxxx= a digit for each node in the cluster indicating the number of CPUs for that node
y= the total database size

For example, *P44DB200* denotes a configuration consisting of 2 nodes each with 4 CPUs switched on and a total database size of 200 GB. The default amount of memory is 2 Gb per node. We will use this notation in the rest of the paper.

4   Experiment Set-up

In this paragraph we continue with a description of the experiments that were performed to investigate the technology. In the experiments we will address manageability, robustness and scalability aspects for the layers Hardware, Operation System and DBMS. The manageability and robustness experiments will have a qualitative character and scalability has a more quantitative character.

4.1   Manageability and Robustness

Manageability is defined as the ease of which the total system is configured and changed. Manageability addresses the issues of configuration, loading, backup and change management. Robustness is defined as how the system handles failures. For example, at the DBMS level this includes recovery from failures of data partitions. At the OS level this includes e.g. recovery from operating system crashes. At the hardware level this includes recovery from nodes or network errors.

4.2   Scalability

Scalability is expressed by three definitions: *scale-up, speed-up* and *workload speed-down*. Scale-up means that we scale the hardware and database proportionally. The scale-up is linear if N times larger hardware and an N times larger database size result in query response times remaining constant. Speed-up is measured by expanding the hardware while keeping the database size constant. The system speed-up is called *linear*, if an N-times larger system results in an N times smaller response time of the workload. Workload speed-down includes experiments with constant hardware configurations, but with an increased database size. Here linear behaviour shows itself by a proportional increase of query response time with the size of the data volume.

Scale-up
The scale-up configurations represent the realistic business setting: adding more Nodes, as the business requires more processing power (increase in database size). Each step the number of nodes is doubled with its full power (four CPU's and 1 GB of memory) and the database size is doubled. These configurations should give a realistic scale-up impression. The configurations for the GBA, GBB, MJ and the MQ experiments used for measuring the scale-up are P4DB100, P44DB200 and P4444DB400.

Speed-up
To test in what perspective the cluster is responsible for this scale-up behaviour, we have investigated system speed-up. The speed-up experiments were performed by increasing system power (1, 2, 4 Nodes) at the database size DB100. Note that the amount of memory as well as the amount of CPU's increases with the number of Nodes. The configurations used for measuring the system speed-up are P4DB100, P44DB100 and P4444DB100.

Workload Speed-down
Now we have investigated the speed-up behaviour, we analyse the scalability behaviour of the queries. This way we can identify to what extent the used algorithms of DB2 UDB EEE are responsible for the scale-up behaviour. The hardware configurations in the different settings are the same. The size of the database varies. The different configurations used for measuring the workload speed-down are P4444DB100, P4444DB200 and P4444DB400.

Data Co-location
As described in the section "Database configuration", the data allocation strategy is very important in a shared-nothing environment. To test its influence, we defined two experiments called *co-located* and *non-colocated*. In both experiments, the MJ query is used as it joins the tables CDR and IPT (CDR.A_NUMBER=IPT.A_NUMBER). In a P4444DB400 configuration, both tables are hashed over all nodes using the their A_Number fields.

In the co-location experiment, the set of A_number values for both the CDR and IPT tables is the same. In the non-colocation experiment, the subsets of IPT data are shifted one node. Fig. 4 shows the data allocation strategies.

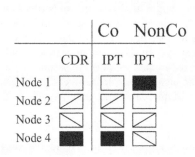

Fig. 4. (Non)co-location data distribution

5   Results of the Experiments

After the above description of the experiments and their context, it is now time to describe the results and conclusions from carrying out the experiments. These results are grouped according to the aspects to be investigated.

5.1   Manageability

At the hardware and operating system level, we made the following observations:
- All machines in the cluster have to be managed separately resulting in extra maintenance effort and increased machine configuration mistakes.
- The Windows NT GUI is easy and convenient for incidental interactive operations but support for regular batch-wise operations is poor.
- Deficiency of people with in-depth Windows NT knowledge for business critical high-end server applications is a high risk for the operations of these kind of Windows NT configurations.

At the database level, we made the following observations:
- The system can almost be managed like a regular one-node database because DB2 provides a single image of the database. The main difference is the data distribution over the available nodes, which requires extra attention.
- the available nodes, which requires extra attention.
- In the current version of DB2 data can be hashed over one or more nodes in the cluster. In some cases however, another distribution based on e.g. manageability criteria may be preferable.

- Both interactive and batch-wise tools are available to manage the database. The interactive graphical database management tool (Control Center) we had, did not show the correct information about the status of the database at all times.
- The 4Gb-per-node log space limitation of DB2[2] is too restrictive especially in a cluster with few nodes and a large amount of data. Both tables residing on one node as multi-node tables are limited in their bulk update or redistribution operations. For data models with a very large fact table, manageability could be improved by allowing maintenance operations (e.g. redistribute, reorganise) on parts of the table at a time (e.g. a table partition or even a sub-partition), which is not supported yet.

5.2 Robustness

At the hardware level, we experienced a lot of problems at the start of the project, e.g. bad memory modules, broken disks, rejected disks because of wrong firmware, bad switch firmware. This frequently resulted in total system crashes. Hence, the first impressions were not very positive. During the project nevertheless, all problems have been solved and from there on, the hardware proved to be reliable.

At the operating system level, the system monitoring facility (Datalog) caused the most problems. Although the primary application (viz. the database) was not directly affected by the problems with this monitoring application, we feel that this is a weakness of the OS. A tight relation between application software (DB2) and the OS (Datalog) should not result in unrecoverable errors (Dr. Watson and rebooting the machine) but in understandable error messages.

At the DBMS level, we experienced that DB2 is a very robust DBMS. We had some initial problems, mainly related to the fact that we were the first to explore the limits of this configuration, but these problems have been solved directly by the development team of DB2. The encountered problems have never led to data integrity violations.

5.3 Scalability

To determine the three scalability factors (scale-up, speedup and speed-down) the response times of the queries described in the case description are used.

Scale-up
The experiment described for scale-up in the section *experiment set-up* resulted in Fig. 5 where we see that the GBA and MJ queries have an above linear scale-up factor. The GBB, which also uses the CDR-table, has a linear scale-up factor and the MQ, which have a relative low response time comparing to the rest of the workload experiments (a couple of minutes instead of one or two hours), have a poor scale-up.

---

[2] In DB2 version 7, the log space per node is limited to 32 Gb.

The cluster handles joins between a large table, such as the CDR table, and a relative small table well. The scalability of this type of workload is excellent as can be seen in the MJ workload experiment. However for MQ, where only relative small tables are joined, scale-up is poor. Here adding extra nodes to the system causes extra overhead, which results in lower response times. Concluding, the setting scales well in a data warehouse environment with one large fact-table and different small dimensions.

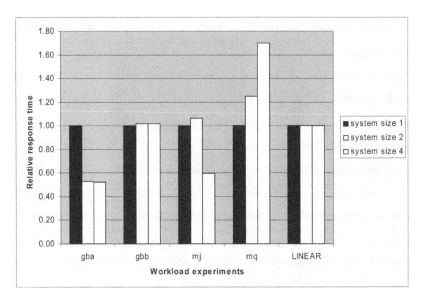

Fig. 5. Scale-up

Speed-up
To measure the speedup we use the response time speedup factor, which is the *inverted* relative response time of the workload queries. The speed-up is *linear* if the *response time speedup* is linear as shown in Fig. 6 for LINEAR. The speed-up is *above linear* if the response time speedup is above linear and the speed-up is *below linear* if the response time speedup is below linear. Hence, the higher the response time speed-up, the better.

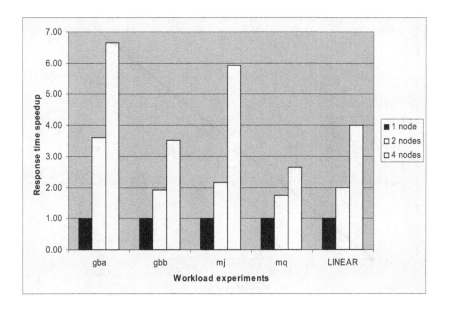

Fig. 6. Speedup

We see that the GBA and MJ, which also had a good scale-up factor, have an above linear response time speedup. The GBB and MQ have a slightly below linear response time speedup. Again the queries which use the large CDR-table scale well.

To figure out what exactly causes this speed-up behaviour, we had a closer look at the following parameters:

- CPU
- Memory
- Switch & I/O system

To do clean experiments regarding the influence of the I/O parameter, we had to vary only this parameter and leave all the other parameters constant. Because the limited number of I/O cards in our configuration we were not able to do these experiments. Instead of testing the I/O influence on the speed-up sec, we investigated the influence of the I/O parameter in combination with the switch.

Effect of CPU
To investigate the effect of the CPU, we have performed experiments with one node, 100GB-database size, constant memory of 2 GB, and varying the number of CPU's from one to four (i.e. P1DB100, P2DB100, P3DB100, P4DB100).

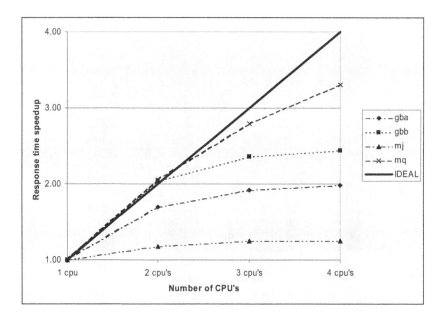

Fig. 7. CPU speedup at DB100

From Fig. 7 we see that the MQ has the most linear CPU response time speed-up. This query appeared to be CPU bound for the one CPU setting. Adding more CPU's, the I/O throughput grows linear to the number of CPU's added. The communication overhead between the CPUs causes the difference between the ideal linear speed-up and what was measured.

CPU is a very important resource for sorting operations. Hence the scalability of GB queries is very sensitive to CPU resources. For the GBB query we see a linear CPU response time speed-up from one CPU to two CPU's. Here the query is CPU bound and the CPU utilisation is above 95%. The GBB becomes I/O bound if more CPU's are added. The GBA has the same characteristic, only this query becomes I/O bound at two CPU's, because of the writing to TEMP. The MJ is already I/O bound using one CPU.

Effect of Memory

The effect of memory should be beneficial for the response time, as larger memory means more efficient in-memory processing. More memory may reduce the need to write data to TEMP space on disk. For this experiment we used the 400GB database setting on 4 nodes and varied the total amount of memory between 4GB and 8GB. In both settings, queries that involved joins were performed with hash joins.

From our measurements however, we saw that the response time using 4GB of memory, for the queries on the large CDR table (GB and MJ), is equal to the response time using 8GB and therefore has no influence on the scaling factor. Even for the marketing query (MQ), which uses less data, the memory is no significant scaling factor. The amount of extra memory (4GB) is to small comparing to the total amount of data processed.

Effect of the Switch
To investigate the effect of the switch we monitored the amount of traffic over the switch on the above described speed-up configurations. The average switch usage is depicted in Fig. 8.

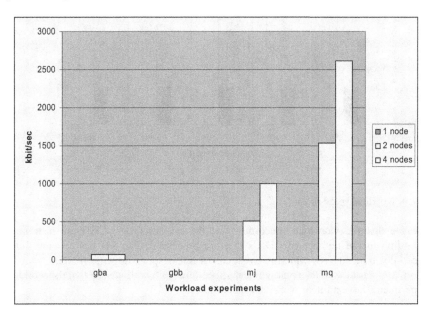

Fig. 8. Average switch usage

The DB2 algorithms are designed in such a way that most of the processing takes place locally on each node, minimising the switch traffic. Considering that the practical maximum throughput of a 10 Mbit hub is around 8 Mbit/s, this should have been sufficient. A commonly used 100 Mbit hub is definitely more than sufficient. Only for the relative small MQ query a 10 Mbit hub is not enough. Here the average switch usage grows linearly when adding more nodes. At 10 nodes the maximum throughput is achieved.

Workload Speed-down
In Fig. 9 the response time is given per workload experiment relative to the response time at the single node setting. The workload speed-down is linear if the relative response time doubles if the amount of data doubles. The workload speed-down is above linear if the relative response time is smaller then the relative database size.

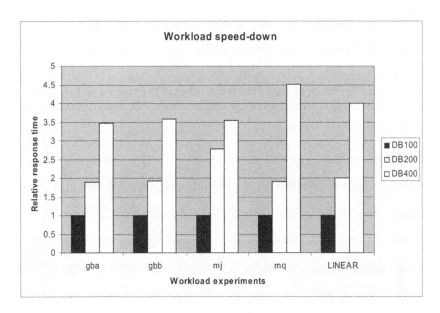

Fig. 9. Workload speed-down

We see that the workload speed-down of the queries GBA, GBB are near linear regarding to the number of CDR's that are processed. The MJ has a below linear speed-down at 200GB, which is worse and a linear speed-down at 400GB, which is good. MQ has at 400GB a below linear speed-down. Overall we see that the workload experiments have a linear workload speed-down.

To explain the non-linear behaviour of the MJ query at 200 GB, we had a closer look at the measurements. First we observed that the TEMP usage of the MJ query increased three times while the database only doubled from 100GB to 200GB. Due to these extra write actions the response time becomes larger. However, when scaling the database from 200 GB to 400 GB we again see a better relative response time than expected. The reason for the efficiency improvement comes from a change of query plan from the DBMS. From Fig. 10 we see that the query plan changed when the data is doubled from 200 GB to 400 GB. The tables LEVERINGEN (LEV) and INDIVIDUEEL_PRODUCT (IPT) are joined by a nested-loop join (NL) at 200 GB and by a hash-join (HJ) at 400 GB. Also the order of the joins changed for the 400 GB. At 200 GB the very large CDR table is joined with the result of the nested loop join, the large result of this join has to be joined to the KLT table and ADS table, which results in a lot of data processing. For 400 GB first all the small tables are joined to each other and finally this result is joined to the large CDR table. The strategy at 400 GB results in less data processing and therefore relative better response time at DB400 and a good scale-up factor.

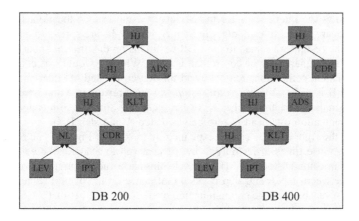

Fig. 10. Query plans for MJ

Data Co-location

We expected the response time of the non-co-location experiments to be significant longer than the equivalent co-location experiments but the co-location of the IPT table had no influence on the response time of the workload experiments.

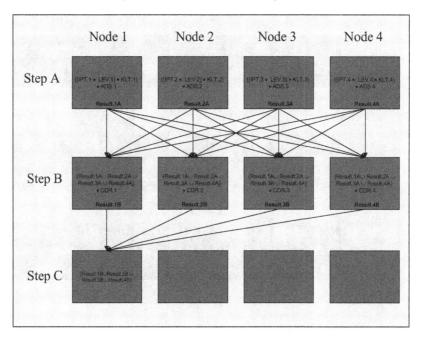

Fig. 11. Query steps for the MJ using non-co-location

Fig. 11 shows the query steps for the MJ query using non-co-location. First locally small tables are joined to each other in Step A, the results are spread to all the nodes. In Step B the results are joined to the CDR partitions on the different nodes. In step C the results are gathered on the co-ordinating node. What we would expect and also see in Fig. 11 is that communication overhead for non-co-located MJ query between step A and step B is *((N-1)/N)*size (total result.A)*, which results in a non-linear speed-up factor for small N and approaches N for large clusters. For the co-location MJ-query this communication overhead is zero between step A and step B, because the data is already on the right node as can be explained by looking at Fig. 12. The results show that the response times differ only a few percent (up to 8%) for the co-located MJ query and the non-co-located MJ query. In this case the communication overhead expected between step A and step B has no influence on the overall response time of the query. A reason why the communication overhead is small is that the SP-switch is fast (300Mbit/sec) and can send the different data streams from Step A to Step B simultaneously.

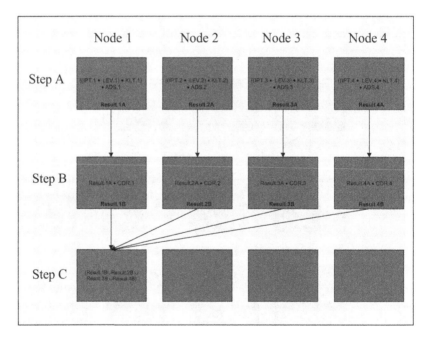

Fig. 12. Query steps for the MJ using co-location

But a more detailed looked at the communication over the switch showed a maximum throughput below 10Mbit/sec, which is far below the practical maximum of 300Mbit/sec of the switch. So for this workload, in a four-node cluster, the capacity of the switch is overkill. From this we can conclude that a database with only one large table and further only smaller tables, little performance gain can be achieved by co-locating the data. Therefore, in some of these cases it might be beneficial to non-co-locate the data, which can result is slightly worse response time, but where the data is easier to manage. If there are two large tables in the data warehouse and consequently a lot of communication between step A and step B, it is advisable to co-locate these tables.

6 Conclusions

Manageability of a cluster system has two viewpoints: the System and the Database. From the database viewpoint, after some start-up problems DB2 provided good manageability despite the fact that there is always room for improvement (e.g. horizontal range partitioning). From the system viewpoint, the manageability of an NT cluster is fair, as each machine has to be managed separately. This requires extra time and increases the chance of (human) errors especially as the number of nodes in the cluster increases.

Robustness of the cluster system is fair. Again after some start-up problems, both at system level and database level, the system could recover from occurring errors without losing data. However, the standard, relative low cost, hardware components appeared to be vulnerable for failures. Especially with an expansion of the cluster, the extra hardware will increase the chance of failures of components. Keep in mind however, that the cluster was not designed for High Availability. Failure of a single node will influence the whole system.

The overall scalability of the cluster system is very good. The workload speed-down shows that the DB2 algorithms scale very well with the growth of data. In a data warehouse environment, one can keep pace with a growing amount of data by adding more hardware. Adding more hardware will increase both processing power as I/O throughput without introducing too much communication overhead. The very fast Netfinity SP switch we used proved to be overkill for this configuration. DB2 is designed in such a way that a standard 100 Mbits/sec hub would suffice. DB2 handles (non) co-location of data very efficiently. In a data warehouse environment with one large fact table and several relative small related tables, one can concentrate on co-location from a manageability point of view rather than from a performance point of view. Furthermore, if the network is fast enough, replication of tables, as we did, might be unnecessary because DB2 can determine to copy small tables on the fly during query execution. If, on the other hand, there are several equally sized large tables, co-location becomes more important with respect to performance .

Although thorough DB2 UDB EEE and Windows NT knowledge is generally available, knowledge of the combination DB2 UDB EEE / Windows NT appeared to be sparsely available. At future deployment time, this aspect has to be re-examined.

Acknowledgements

We would like to thank the DB2 team in Toronto for their timely and extensive support. In particular we would like to thank Berni Schiefer and Alex Hazlitt. Without their help we would not have been able to perform our experiments.

References

1. EURESCOM, Project   P817, "Database technologies for large scale databases in Telecommunications", Deliverable 1, Overview of Very Large Database Technologies and Telecommunication Applications using such Databases (March 1999)
2. Derks, W.L.A. et. al., "Experimenting NUMA for Scaleable CDR Processing", in: Proceedings of DEXA 2000, London, Greenwich, September 2000.
3. Schiefer, B. et. al., "IBM's DB2 Universal Database demonstrations at VLDB '98", in: Proceedings of the 24th annual international conference on Very Large Data Bases, NY, August 1998
4. Compaq(Tandem), "NonStop SQL/MX Demonstration: Two-Terabyte Database on Windows NT Server", white paper at http://servernet.himalaya.compaq.com/flat/public/brfs_wps/smx2tbbf/smx2tbbf.htm, May 1997

Mining Sequential Alarm Patterns in a Telecommunication Database

Pei-Hsin Wu, Wen-Chih Peng and Ming-Syan Chen

Department of Electrical Engineering,
National Taiwan University
Taipei, Taiwan, ROC
{mschen@cc.ee.ntu.edu.tw, wcpeng,peggywu@arbor.ee.ntu.edu.tw}

Abstract. A telecommunication system produces daily a large amount of alarm data which contains hidden valuable information about the system behavior. The knowledge discovered from alarm data can be used in finding problems in networks and possibly in predicting severe faults. In this paper, we devise a solution procedure for mining sequential alarm patterns from the alarm data of a GSM system. First, by observing the features of the alarm data, we develop operations for data cleaning. Then, we transform the alarm data into a set of alarm sequences. Note that the consecutive alarm events exist in the alarm sequences, and it is complicated to count the occurrence counts of events and extract patterns. Hence, we devise a new procedure to determine the occurrence count of the sequential alarm patterns in accordance with the nature of alarms. By utilizing time constraints to restrict the time difference between two alarm events, we devise a mining algorithm to discover useful sequential alarm patterns. The proposed mining algorithm is implemented and applied to test against a set of real alarm data provided by a cellular phone company. The quality of knowledge discovered is evaluated. The experimental results show that the proposed operations of data cleaning are able to improve the execution of our mining algorithm significantly and the knowledge obtained from the alarm data is very useful from the perspective of network operators for alarm prediction and alarm control.

1 Introduction

Due to recent technology advances, an increasing number of users are accessing various information system via wireless communication. Such information systems as stock trading, banking, wireless conferencing, are being provided by information services and application providers [5][6][7][13], and mobile users are able to access such information via wireless communication from anywhere at anytime [3][12][15].

W. Jonker (Ed.): Databases in Telecommunications II, LNCS 2209, pp. 37-51, 2001.

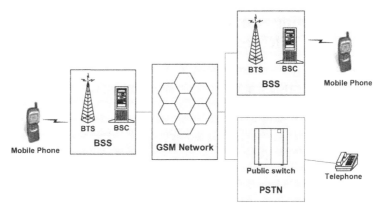

Fig. 1. The general architecture of a GSM system.

The architecture of wireless networks varies from one standard to another. Use of the *Global System for Mobile Communication* (GSM) continues to spread throughout the world. Figure 1 shows the general architecture of the GSM system. It can be seen that a GSM system comprises lots of interconnected components and each component also contains several subcomponents. Individual components and subcomponents generate alarms indicating some sort of abnormal situations. Table 1 shows an example of selected real alarm data where sourceID is the identification of an individual component that generates the alarm and errorID indicates the abnormal situation. Thus, a GSM system produces daily a large amount of alarm data which contains hidden valuable information about the system behavior. The knowledge discovered from alarm data can be used in finding problems in networks and possibly in predicting severe faults. Since the number of alarms could be in tens of thousands daily, it is infeasible to investigate these alarms manually. In order to extract the valuable knowledge from the alarm data, we utilize the techniques of data mining in this study.

Table 1. An example of selected alarm data.

day	time	Source	sourceID	errorID	...
2001/01/01	15:07:09	SITE	21233	a	...
2001/01/01	15:46:23	SITE	21233	f	...
2001/01/01	19:04:37	BSC	bsc05	z	...
2001/01/01	22:15:44	SITE	33009	h	...
2001/01/01	23:20:15	SITE	10021	q	...
...	...	...	...	...	...

Data mining, which is also referred to as *knowledge discovery* in databases, means a process of extracting implicit, previously unknown and potentially useful information (in terms of knowledge rules, constraints, regularities) from data in databases. By knowledge discovery in databases, interesting knowledge, regularities, or high-level information can be extracted from the relevant sets of data in databases and be investigated from different angles. Various data mining capabilities have been explored in the literature [1][2][4][8][9][10][16]. It is noted that utilizing the technique of mining sequential patterns is able to extract valuable knowledge from the alarm data generated by a GSM system [11][14]. In mining sequential patterns, the input data is a set of sequences, called *data-sequences*. Each data-sequence is a list of transactions, where each transaction is a set of literals, called *items*. Typically there is a transaction-time associated with each transaction. A *sequential pattern* also consists of a list of sets of items. The problem is to find all sequential patterns with a user-specified minimum *support*, where the support of a sequential pattern is the percentage of data-sequences that contain the pattern. In prior works, the constraint imposed by the time difference between two consecutive events has not been explored. For example, the customers typically rent "Star Wars" first, and then "Empire Strikes Back". Note that such a pattern did not contain the time difference of these two renting activities. Hence, the knowledge of time difference has not been taken into account when the mining results are produced. However, such a time difference between two events is in fact an important knowledge when mining the alarm data since, with this knowledge, one can not only judge the relevance of these two events, but also predict the alarm sequence and take proper steps to prevent the occurrence of the alarms if all possible.

An example sequential alarm sequence that we shall explore in the alarm data is shown in Figure 2. The number in each circle represents the errorID, same as in Table 1, and $T_{i,j}$ denotes the time difference between alarm event$_i$ and alarm event$_j$. As mentioned above, this knowledge is important from the perspective of network operators in that a network operator can hence be warned beforehand of severe alarms and take proper provision. For example, if the network operator detects that the alarm a occurring at time t, he/she should dissipate this alarm before the time $t + T_{a,q}$ to alleviate the abnormal situations incurred. It can be seen that the problem of mining sequential patterns for the telecommunications is intrinsically different from that for the rental industry, and particularly ought to capture the presence of time difference between events. Therefore, it is necessary to develop a new mining algorithm for mining constrained sequential alarm patterns for a telecommunication system.

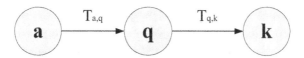

Fig. 2. An example of sequential alarm patterns.

In this paper, we devise a solution procedure for mining sequential alarm patterns from the alarm data of a GSM system. First, by observing the features of the alarm data, we develop two operations for data cleaning. Specifically, one operation is to delete those undetermined alarm events and the other is to merge the repeated alarm events caused by the same abnormal situation. After the data cleaning procedure, we transform the alarm data into a set of alarm sequences. Note that the consecutive alarm events exist in the alarm sequences, and it is complicated to count the occurrence counts of events and extract meaningful alarm patterns. Hence, we devise a new counting method to determine the occurrence count of the sequential alarm patterns in accordance with the nature of alarms. By utilizing time constraints to restrict the time difference between two alarm events, we devise an algorithm of mining sequential alarm patterns (to be referred to as algorithm MSAP) to discover useful sequential alarm patterns. The algorithm MSAP is implemented and applied to test against a set of real alarm data provided by a cellular phone company in Taiwan. The quality of knowledge discovered is evaluated. The experimental results show that the proposed operations of data cleaning are able to improve the execution of our mining algorithm significantly and the knowledge obtained from the alarm data is very important from the perspective of network operators for alarm prediction and alarm control.

This paper is organized as follows. Preliminaries are given in Section 2. In Section 3, we devise the constraint-based algorithm for mining sequential alarm patterns. Experimental results are presented in Section 4. This paper concludes with Section 5.

2 Preliminaries

To facilitate the presentation of this paper, some preliminaries are given in this section. In Section 2.1, we describe the alarm data in a GSM system. In Section 2.2, the procedure of mining sequential alarm patterns is presented.

2.1 Alarm Data in GSM

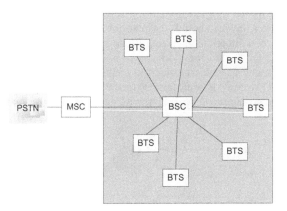

Fig. 3. The components of a GSM system: MSC, BSC, and BTS.

Figure 3 shows the components of a GSM system. The Mobile Services Switching Center (MSC) is a complete exchange system which is able to route calls from the fixed network to an individual mobile station. The MSC has connections to other entities of the GSM system. Those connections allow the MSC to gather the current statuses of mobile stations. The Base Station Controller (BSC) provides the management of the Base Transceiver Station (BTS) and also the communication between the MSC and the BTS. The BTS supports the radio transmission/reception and the control of radio functions.

When detecting an anomalous situation, the GSM system generates an alarm event to alert the operators for this anomalous situation. All these alarm events are recorded in the alarm log table, which has several attributes, such as day, time, source, sourceID, severity and errorID. The attributes *day* and *time* record the time when the alarm event occurred. The attribute *source* is the type of the component generating the alarm event. There are three types of sources, cell, site, and BSC. Explicitly, a cell refers to a cell inside a BTS, a site refers to a base station and BSC is the base station controller. As to the other attributes of alarm log table, *sourceID* is the identification of an individual component and *severity* means the severity of the anomalous situation. Each anomalous situation or alarm event is associated with the identification number, denoted as *errorID*. For example, errorID "m" indicates that "unavailable radio signaling channels threshold critical" occurs.

2.2 Procedure of Mining Alarm Sequential Patterns

As described before, a GSM system produces daily large amount of alarm data in the alarm log data. In this paper, we explore the problem of mining sequential alarm patterns from the alarm log data. The procedure of mining sequential alarm patterns that involves the following tasks:

1. Data cleaning: In order to extract useful sequential alarm patterns, data cleaning and data preprocessing are needed [10]. The operations of data cleaning include selecting those attributes needed for data mining and removing those redundant information or outliers.

2. Discovery of patterns: After the process of data cleaning, we employ a data mining algorithm to discover the valuable sequential alarm patterns from the alarm data. The sequential alarm patterns contain not only the temporal relationship but also the time spans among the alarm events.

With the problem and the corresponding tasks described above, we then develop an approach to mining sequential alarm patterns.

3 Mining Sequential Alarm Patterns

In this section, we describe the process of data cleaning in Section 3.1. In Section 3.2, we describe the algorithm of mining sequential alarm patterns (MSAP) to extract useful sequential alarm patterns.

3.1 Data Cleaning

The goal of data cleaning is to identify the available data sources and extract the data that is needed for preliminary analysis in preparation for future mining. Clearly, the process of data cleaning depends on the patterns to be mined. Note that the alarm log data contains not only useful information but also dummy information. Before mining sequential alarm patterns, the procedure of data cleaning is needed. Specifically, we have the following two operations for data cleaning:

Cleaning undetermined alarm events

Some anomalous causes that cannot be recognized by the alarm system are assigned their errorID to be 88888 and the alarm event with its errorID 88888 is frequently found in the alarm log data. Becasue of the frequent occurrence of the alarm event 88888, the sequential alarm patterns are those patterns containing lots of the alarm event 88888. Clearly, these sequential alarm patterns are of less interest. Thus, we shall clean undetermined alarm events to improve the quality of the sequential alarm patterns mined.

Merging repeated alarm events

It is also noted that some repeated alarms appear frequently in the alarm log data. It can be verified that some repeated alarms are independent events and the others are caused by the same abnormal situations. For example, the alarm log data has the sequence <AAAABC> in which alarm event A repeats four times. Notice that if these repeated alarm events are from the same abnormal situation, we would like to merge them to one alarm event (i.e., the <AAAABC> becomes <ABC>). Otherwise, we retain the repeated alarms. Note that not all repeated alarm events could be merged. Merging the repeated alarm events should be done in accordance with the *ceased table* in the telecommunication database system. The ceased table contains the time when an alarm event is generated and the time when the alarm event is ceased. Table 2 shows an example of a ceased table, where Dtime is the generated time of an alarm, Ctime is the ceased time of this alarm and Duration is the duration between Dtime and Ctime in seconds.

Table 2. An example of a ceased table.

Dtime	Ctime	duration (in seconds)	source	sourceID	errorID
2000/09/04 15:15:27	2000/09/06 00:03:08	18061	BSC	bsc22	q
2000/09/04 16:14:54	2000/09/09 16:55:52	434458	SITE	40833	s
2000/09/04 16:42:13	2000/09/09 11:37:15	413702	SITE	23004	c
2000/09/04 17:41:08	2000/09/08 09:22:04	333656	CELL	51821	i

3.2 Algorithm MSAP: Mining Sequential Alarm Patterns

After the data cleaning procedure, we transform the alarm data into a set of alarm sequences. Then, we devise an algorithm MSAP (standing for Mining Sequential Alarm Patterns) to extract useful sequential alarm sequences. Denote the set of errorID as E and define an alarm event as a pair (eid, t), where $eid \in E$ is an errorID and t is the time when the alarm event occurred. For example, an alarm event (A, 2001/2/10 16:14:08) means that the alarm A occurred at 16:14:08 on 10 Feb in 2001.

Definition 1: An alarm sequence S_{sid} is a sequence of ordered alarm events generated by the component sid, where sid is the sourceID and $S_{sid} = <(eid_1,t_1), (eid_2,t_2), (eid_3,t_3),..., (eid_n,t_n)>$, where $t_i < t_j$ if $i < j$.

As described before, the alarm events are generated form individual components of the GSM. In this paper, we address the problem of mining sequential alarm patterns for the individual components and create an alarm sequence table containing alarm sequences. Table 3 is an alarm sequence table where the component of GSM is site (i.e., the base station) and the duration time of the alarm sequences is two hours, from 0:00am to 2:00am. Four alarm sequences (i.e., S_{2001} , S_{2002}, S_{2003} and S_{2004}) are given in Table 3. The *length* of an alarm sequence S_{sid} , denoted as $| S_{sid} |$, is the number of events in the sequence sid. A sequence of length k is called a k-sequence. It can be seen that the length of S_{2001} is four.

Table 3. An example of alarm sequences.

Site ID	Alarm sequences
2001	< (A, 0:12) (C, 0:25) (D, 1:07) (C, 1:19) >
2002	< (C, 0:37) (C, 0:42) (B, 1:11) (A, 1:36) >
2003	< (B, 0:05) (C, 0:21) (C, 0:54) (A, 1:25) >
2004	< (B, 0:27) (C, 0:37) (A, 1:22) >

The time difference between two events is in fact an important knowledge when mining the alarm data since, with this knowledge, one can predict the alarm sequence and take proper steps to prevent the occurrence of the alarms if all possible. In order to find the critical sequential alarm patterns concerned by network operators, algorithm MSAP allows network operators to set the restriction on the time difference between two alarm events in sequential alarm patterns. Thus, we have the following definition.

Definition 2:*Urgent window*, denoted as δ, is a user-specified time interval. We only count the pattern p that matches the candidate pattern $c \in C_2$, if the alarm events of p occur within the interval of an urgent window δ.

It will be seen that the size of an urgent window is set to determine the maximal time lag two events are allowed if they are to be viewed relevant to each other. With the set of the alarm sequences obtained, we next determine the sequential alarm patterns. Algorithm MSAP is devised and applied to alarm sequences to discover sequential alarm patterns. C_k is a set of candidate k-sequences. The occurrence count of C_k in alarm sequences is referred to as support. Let L_k represent the set of large k-sequences, where there are a sufficient number of alarm sequences containing this k-sequence. Such a threshold number is called minimum support in this paper. Using a candidate generation procedure, algorithm MSAP makes multiple scans over the set of alarm sequences to generate the possible sequential alarm patterns.

By observing the alarm sequences, the processing of counting support is nontrivial. For the customer transaction data, the support is usually only increased by one per customer even if the customer buys the same set of items in two different transactions. In contrast, alarm event A and alarm event C could occur several times in one day. As a consequence, it is essential to develop the counting method to obtain the support of these candidate alarm sequences.

In order to find useful sequential alarm patterns, we only consider the event pair (i, j), where i and j are alarm events and alarm j is the closest to alarm event i.

Fig. 4. An example of an alarm sequence.

Figure 4 shows the example of counting the support of <AC>. Note that when we find an alarm event A, we only include for counting the alarm event C that is the closest to the alarm event A. Therefore, <(A, 0:12) (C, 1:42)>, <(A, 0:12) (C, 1:51)>, <(A, 0:38) (C, 1:51), and <(A, 1:25) (C, 1:51)> are not counted by algorithm MSAP. Thus, in this example, the support of <AC> is three as shown in Figure 5.

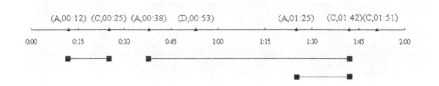

Fig. 5. The occurrence count of the sequential alarm sequence <AC>.

With the above counting process, the large k-sequence has the following property:

Property 1: For any sequential alarm pattern $pattern_j = <r_1, r_2, ..., r_k>$ in L_k, the support of this sequential alarm sequence is smaller than that of the sequential alarm patterns $pattern_j = <q_1, q_2, ..., q_{k-1}>$ in L_{k-1}, where $q_x = r_x$ and $1 \le x \le k-1$. For example, the support of <ABDF> is smaller than that of <ABD>.

In light of Property 1, we devise *Procedure Candidate_Generation* to generate the possible candidate k-sequences and algorithm MSAP to mine sequential alarm patterns. This solution procedure is outlined as follows.

Procedure Candidate_Generation
insert into C_k
select $p.event_1, p.event_2, ..., p.event_{k-1}, q.event_1$
from $p \in L_{k-1}, q \in C_1$

Algorithm MSAP
Input: A set of alarm sequences $\{S\}$, urgent window δ, minimum support *sup*
Output: sequential alarm patterns
1. $L_1 = \{$large 1-sequence$\}$
2. $C_2 =$new candidates generated from L_1 by calling
        Procedure Candidate_Generation
3. **for** each alarm-sequence s in the database **do**

4.   **for** each candidates $c \in C_2$ **do**

5.   search c in the sequence s

6.   **if** an event pair p matched $c \in C_2$ is found and the time-difference of event pair p is less than the *urgent window* δ

7.   increase the count of $c \in C_2$ and record the time-difference

8. L_2=candidates in C_2 with minimum support required

9.   Compute the mean μ and standard deviation σ of the time-difference of each $l \in L_2$

10. **for** (k=3 ; $L_{k-1} \neq 0$; k++) **do**

11.   **begin**

        C_k= new candidates generated from L_{k-1} by calling *Procedure Candidate_Generation*

12.   **for** each alarm-sequence s in the database **do**

13.   **for** each candidates $c \in C_k$ **do**

14.   search c in sequence s

15.   **if** a pattern x matched c is found and the time-difference of each event is less than or equal to δ

            increase the count of the candidate $c \in C_k$

16. L_k=candidates in C_2 with minimum support required

17.   **end**

Note that L_1 can be simply generated from the counting of appearing alarm events. Then, C_2 are obtained by *Procedure Candidate_Generation* (line 2 of algorithm MSAP). After generating possible C_2, the occurrence count and the time difference of all candidate $c \in C_2$ are obtained by scanning all the alarm sequences (from line 3 to line 7 in algorithm MSAP). Algorithm MSAP calculates the mean and standard deviation of time-difference for each $l \in L_2$ (line 9 of algorithm MSAP). L_2 can be determined by proper trimming on C_2. By utilizing *Procedure Candidate_Generation*, C_k can be generated by $L_{k-1}*C_1$ where * is an operation for concatenation (from line 10 to line 18 in algorithm MSAP), which is very different from the one in the conventional sequential pattern mining. The support of each k-sequences is determined for the identification of L_k.

3.3 An Illustrative Example for Mining Sequential Alarm Patterns

Consider an example alarm sequence given in Table 3. In each iteration (or each pass), algorithm MSAP constructs a candidate set of large sequences, counts the number of occurrences of each candidate sequence, and then determines large sequences based on a pre-determined minimum support. In the first iteration, algorithm MSAP simply scans all the alarm sequences to count the number of occurrences for each alarm event. The set of candidate 1-sequence, C_1, obtained is shown in Table 4a. Assume that the minimum support required is 2. The set of large

1-sequences, L_1, composed of candidate 1-sequences with the minimum support required, can then be determined.

Table 4. Generation of candidate sequences and large sequences

(a) Generation of C_1 and L_1.

C_1	support
A	4
B	3
C	7
D	1

L_1	support
A	4
B	3
C	7

(b) Generation of C_2 and L_2.

C_2	support	time difference	μ	σ
AA	0		0	0
AB	0		0	0
AC	1	13	13	0
AD	1	55	55	0
BA	2	25, 55	40	15
	0		0	0
BC	2	16, 10	13	3
BD	0		0	0
CA	4	59, 54, 31, 45	47.25	10.64
CB	2	34, 29	31.5	2.5
CC	3	54, 5, 33	30.67	20.07
CD	1	42	42	0

L_2	support
BA	2
BC	2
CA	4
CB	2
CC	3

(c)   Generation of C_3 and L_3.

C_3	support
BAA	0
BAB	0
BAC	0
BAD	0
BCA	1
BCB	0
BCC	1
BCD	0

C_3	support
CAA	0
CAB	0
CAC	0
CAD	0
CBA	2
CBB	0
CBC	0
CBD	0

C_3	support
CCA	2
CCB	1
CCC	0
CCD	0

L_3	support
CBA	2
CCA	2

To discover the set of large 2-sequence, algorithm MSAP uses L_1*C_1 to generate a candidate set of sequences C_2. Next, the four alarm sequences in Table 3 are scanned. The support of each candidate sequence in C_2 whose time difference between two alarm events is smaller than δ (i.e., the urgent window) is counted. Table 4b represents the result from such counting in C_2. As described before, time plays an important role in mining sequential alarm sequences. Thus, algorithm MSAP calculates the mean and the standard deviation of the time difference for each 2-sequence in C_2. The set of large 2-sequence, L_2, is therefore determined based on the support of each candidate 2-sequence in C_2. The set of candidate 3-sequence is generated from $L_2 * C_1$. Algorithm MSAP then scans the alarm sequences and discovers the large 3-sequences in Table 4c. Following the same procedure, algorithm MSAP discovers the large k-sequences. If there is no other large k-sequence discovered, algorithm MSAP stops and those large k-sequences are the sequential alarm patterns mined from the alarm data.

4 Experimental Results

The effectiveness of mining sequential alarm sequences is evaluated in this section. The alarm log data are provided by a cellular phone service provider. Experimental results of utilizing data cleaning techniques are shown in Section 4.1. Then, we present the experimental results of algorithm MSAP in Section 4.2.

4.1 The Impact of Data Cleaning

As mentioned before, the alarm log data contains not only useful information but also dummy information, thus calling for data cleaning. The execution time of mining sequential alarm patterns can also be reduced. We now examine the impact of data cleaning. We select one week alarm log data of site from the alarm database, where the number of sites is 3830. The performance improvement of MSAP is computed as follows.

$$\text{performance improvement} = \frac{|\text{execution time before cleaning} - \text{execution time after cleaning}|}{\text{execution time before cleaning}}$$

We then investigate the performance improvement when the operations of data cleaning are applied to the alarm data. Figure 7 shows the performance improvement while performing the procedure of data cleaning in advance. It can be seen that the execution times of algorithm MSAP vary due to the reason that the numbers of alarm events generated are different. Since the undetermined alarm events are substantial, occupying almost one-third of the alarm data volume before cleaning, the curve of the performance improvement ranges from 25% to 50%. For the process of merging repeated alarm events, the amount of repeated alarms is small. It can be seen in Figure 7 that the performance improvement is dominated by the operation of cleaning undetermined alarm events. Notice that the execution of algorithm MSAP after data cleaning is much faster than that without data cleaning, showing the merit of the data cleaning procedure.

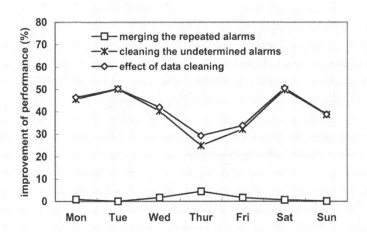

Fig. 7. The performance improvement while performing the operations of data cleaning.

4.2 Experimental Results of Algorithm MSAP

In this section, we evaluate the correctness of the sequential alarm patterns discovered by algorithm MSAP. Without loss of generality, we select one week alarm log data, and set the value of the minimum support to be 2000 and that of the urgent window to be one hour. Table 5 shows the selected sequential alarm patterns mined by algorithm MSAP and the descriptions of errorID is shown in Table 6. As can be seen in Table 5a, the time difference of two alarm events in L_2 is calculated by algorithm MSAP and is very important in predicting the network behaviors. For example, the sequential alarm pattern $<(g)(e)>$ with its mean 815 and variance 962 is interpreted as when the problem of general line occurs, the loss of the frame alignment will likely occur after 815 seconds.

Table 5. The selected sequential alarm patterns discovered by algorithm MSAP.

(a) The selected large 2-sequences (L_2).

L_2	mean μ (in seconds)	standard deviation σ (in seconds)
$<(a)(b)>$	803	593
$<(a)(c)>$	965	944
$<(b)(d)>$	1406	1202
$<(g)(e)>$	815	962
$<(g)(f)>$	1003	978

(b) The large 3-sequences (L_3).

<(a)(b)(a)>
<(a)(b)(c)>

Table 6. Descriptions of errorID.

errorID	Description
a	LPDL link down
b	LPDL link transition
c	Synchronization changed to holdover mode
d	End of holdover mode
e	Loss of frame alignment
f	Lower BER threshold exceeded
g	General line problems
h	Site Input Alarm detected

Table 7. The mining results obtained by algorithm MSAP with the value of the urgent window varied.

(a) The selected L_2 with the value of the urgent window to be one hour.

L_2	mean μ (in seconds)	standard deviation σ (in seconds)
<(b)(b)>	176	462
<(g)(e)>	826	980
<(g)(g)>	105	293
<(g)(f)>	962	946

(b) The selected L_2 with the value of the urgent window to be two hours.

L_2	mean μ (in seconds)	standard deviation σ (in seconds)
<(h)(h)>	480	1185
<(b)(b)>	240	720
<(b)(d)>	634	1308
<(g)(e)>	1813	2117
<(g)(g)>	132	468
<(g)(f)>	1759	1905

The impact of varying the values of the urgent window δ is next investigated. We select one day alarm log data and set the value of the minimal support to be 2000. The alarm sequences obtained are shown in Table 7. It can be seen from Table 7b that more sequential alarm sequences (i.e., <(h)(h)> and <(b)(d)> are discovered by algorithm MSAP when the value of the urgent window becomes larger. This is due to the reason that with a larger value of δ, the support of alarm event pairs increases. The selection of the value of δ is determined by network operators and is also dependent upon the distribution of time difference between alarm event pairs.

It is worth mentioning that the sequential alarm patterns are beneficial for managing and designing alarm systems. With the sequential alarm patterns discovered, one can predict the alarm sequence and take the proper steps to prevent the occurrence of alarms if all possible. Furthermore, as pointed out in [11], alarm filtering requires the knowledge of the alarm sequences. By combining and transforming the related alarm events into one alarm, one is able to design proper alarm systems for telecommunication systems. Clearly, these sequential alarm patterns mined by our algorithm MSAP are also very useful in these design issues.

5   Conclusions

In this paper, we devised a solution procedure for mining sequential alarm patterns from the alarm data of a GSM system. First, by observing the features of the alarm data, we developed operations for data cleaning. After the data cleaning procedure, we transformed the alarm data into a set of alarm sequences. We devised a new counting method to determine the occurrence count of the sequential alarm patterns in accordance with the nature of alarms. By utilizing time constraints to restrict the time difference between two alarm events, we devised algorithm MSAP to obtain the mean and the standard deviation of two alarm events, and discover useful sequential alarm patterns. Algorithm MSAP is implemented and applied to test against a set of real alarm data provided by a cellular phone service provider. The quality of knowledge discovered is evaluated. The experimental results showed that the proposed operations of data cleaning are able to improve the execution of our mining algorithm.

Acknowledgement

The authors are supported in part by the Ministry of Education Project No.89-E-FA06-2-4-7 and the National Science Council, Project No. NSC89-2219-E-002-028 and NSC 89-2218-E-002-028, Taiwan, Republic of China

Reference

1.   R. Agrawal, T. Imielinski, and A. Swami: Mining Associations between Sets of Items in Massive Databases. In Proceedings of ACM SIGMOD (May 1 993) 207—216.
2.   R. Agrawal and R. Srikant. Mining Sequential Patterns. Proceedings of the Eleventh IEEE Inter-national Conference on Data Engineering (1995) 3—14.
3.   B. Bruegge and B. Bennington. Applications of Mobile Computing and Communication. IEEE Personal Communication (February 1996) 64—71.

4. M-S. Chen, J. Han, and P. S. Yu.: Data Mining: An Overview from Database Perspective. IEEE Transactions on Knowledge and Data Engineering (December 1 996) (6):866—883.
5. N. Davies, G. S. Blair, K. Cheverst, and A. Friday: Supporting Collaborative Application in a Het-erogeneous Mobile Environment. Computer Communication Specical Issues on Mobile Computing (1996).
6. M. H. Dunham: Mobile Computing and Databases. Tutorial of International Conference on Data Engineering (February 1998).
7. A. Elmagarmid, J. Jain, and T. Furukawa: Wireless Client/Server Computing for Personal Information Services and Applications. ACM SIGMOD RECORD (December 1995) 24(4):16—21.
8. Minos N. Garofalakis, Rajeev Rastogi, and Kyuseok Shim: SPIRIT: Sequential Pattern Mining with Regular Expression Constraints. In Proceedings of VLDB (September 1999) 223-234.
9. J. Han, G. Dong, and Y. Yin: Efficient Mining of Partial Periodic Patterns in Time Series Database. In Proceedings of the 15th International Conference on Data Engineering (March 1999).
10. J. Han, M. Kamber: Data Mining: Concepts and Techniques. Academic Press (2001).
11. K. Hätönen, M. Klemettinen, H. Mannila, P. Ronkainen, and H. Toivonen: Knowledge discovery from telecommunication network alarm databases. In Proceedings of the Twelfth International Conference on Data Engineering, New Orleans, Louisiana (1996) 115—122.
12. N. Krishnakumar and R. Jain: Escrow Techniques for Mobile Sales and Inventory Applications. ACM Journal of Wireless Network (July 1997) 3(3):235—246.
13. D. L. Lee: Data Management in a Wireless Environment. Tutorial of International Conference on Database System for Advance Applications (April 999).
14. H. Mannila, H. Toivonen, and A. I. Verkamo: Discovery of Frequent Episodes in Event Sequences. Data Mining and Knowledge Discovery (1996) 1(3):259—289.
15. M. Satyanarayanan: Mobile Information Access. IEEE Personal Communication (February 1996) 26—33.
16. Marek Wojciechowski: Interactive Constraint-Based Sequential Pattern Mining, Proc. of the 5th East European Conference on Advances in Databases and Information Systems (ADBIS'01), Vilnius, Lithuania (2001), to appear.

Generalized MD-Joins:
Evaluation and Reduction to SQL

Michael O. Akinde and Michael H. Böhlen

Department of Computer Science, Aalborg University, Denmark
{strategy, boehlen}@cs.auc.dk

Abstract. On-line analytical processing (OLAP) has become an increasingly important concern for telecommunications operators. Telecommunications systems generate huge amounts of data that would be beneficial to analyze using OLAP technology. Recently, Chatziantoniou et al. [4] presented the MD-join, a relational operator that provides a clean separation between group definitions and aggregate computations, and allows to succinctly express OLAP queries. In this paper, we define generalized MD-joins, describe an implementation of the GMD-join query engine on top of a commercial DBMS, and present a reduction of GMD-joins to SQL. We present a practical new optimization of GMD-joins allowing the restriction of base-values, and discuss the optimization of GMD-joins with restrictions and coalescing of GMD-joins. We show how GMD-join optimizations can be used to improve the performance of the translated SQL queries. Finally, we present experiments comparing the GMD-join query engine with the SQL reduction.

1 Introduction

On-line analytical processing (OLAP) has become an increasingly important concern in telecommunications systems operations in recent years. Managing telecommunications network operations often involves debugging network problems, optimizing network configuration, detecting inconsistencies in network data (e.g., for fraud detection), etc., and requiring complex analysis of the network data. Most such analyses are currently implemented by network engineers using complex algorithms coded in procedural programming languages like Perl.

Many of these analyses can be implemented with OLAP queries, using SQL grouping and aggregation, data cubes [8], marginal distributions extracted using the unpivot operator [7], and so on. In fact, using OLAP technology for such purposes can greatly speed up data analysis [1]. Recently, Chatziantoniou et al. [4] presented the MD-join, an aggregation-join operator that can be used to formalize complex OLAP queries such as those often found in telecommunications queries. The MD-join is a flexible and powerful operator that allows to generalize a large range of OLAP query optimizations within the framework of MD-join query optimizations.

W. Jonker (Ed.): Databases in Telecommunications II, LNCS 2209, pp. 52-67, 2001.

In this paper, we report on our work with building a generalized MD-join (GMD) query engine on top of an SQL-based commercial DBMS. We also present a reduction of GMD-joins to SQL that more closely models the evaluation of the GMD-join, and is thus more efficient than prior attempts to model the MD-join using standard query languages. We discuss the optimization of this SQL reduction, and also present an optimization on the GMD-join for restricting the base-values relation. Finally we present experiments comparing the two evaluation methods presented in this paper, and discuss the conflicting optimizations methods of the GMD-join.

1.1 Related Work

A variety of OLAP frameworks have been proposed in the literature, allowing for a great degree of control over the query. Extensions such as Cube By [8] and unpivot [7] allow alternative definitions of groups. EMF-SQL [2, 3, 5] permits the definition of customized aggregate conditions using the definition of grouping variables. Another approach to permit the specification of complex aggregate queries is to allow user-defined aggregate functions (e.g., [9, 10, 11]). The MD-join[4] combines complex group specification with complex aggregate specification in a single relational framework.

The issue of an efficient reduction of generalized MD-joins to SQL has not previously been considered in the literature. Evaluation and performance measurements of EMF-SQL algorithms (broadly similar to the MD-join) are considered in the EMF-SQL literature [2, 3]; however this body of work does not consider the optimization of SQL queries using the underlying MD-join model. Restricting the base-values relation and the interaction between base-values/detail table restrictions and coalescing in GMD-joins are not considered in either the MD-join or EMF-SQL literature.

1.2 Paper Outline

The paper proceeds as follows: In Section 2, we briefly present the GMD-join operator and present decision support examples over flow data. In Section 3, we discuss the computation algorithm for GMD-joins over a commercial DBMS, and give a SQL reduction of GMD-joins that allows for an accurate computation of GMD-joins using SQL. In Section 4, we discuss optimization of GMD-joins, and show how such optimizations can used to make the SQL reduction more efficient. Finally, Section 5 presents the comparison studies and experiments between the GMD-join query engine and the SQL reduction.

2 Preliminaries

The *MD-join operator* is used to algebraically express many complex OLAP queries. It provides a clean separation between the definition of the groups and the definition of aggregates in an OLAP query. The *generalized MD-join operator* is an extension of the MD-join that allows us to reduce a series of nested MD-joins to a single generalized expression [4].

Let θ be a condition, b be a tuple, and R be a relation. We write attr(θ) to denote the set of attributes used in θ. RNG$(b,R,\theta) = \{r \mid r \in R \wedge \theta(b,r)\}$ denotes the set of tuples in R that satisfies θ. We write $\{\{...\}\}$ to denote a multiset.

Definition 2.1
Let $B(\boldsymbol{B})$ and $R(\boldsymbol{R})$ be relations, θ_i be a condition with attr(θ) $\subseteq \boldsymbol{B} \cup \boldsymbol{R}$, and l_i be a list of aggregate functions $(f_{i1}, f_{i2}, ..., f_{in})$ over attributes $c_{i1}, c_{i2}, ..., c_{in}$ in $\boldsymbol{R}$. The GMD-join, MD$(B, R, (l_1,...,l_m), (\theta_1,...,\theta_m))$, is a relation with schema $\boldsymbol{X} = (\boldsymbol{B}, f_{11}\_R\_c_{11}, ..., f_{1n}\_R\_c_{1n}, ..., f_{m1}\_R\_c_{m1}, ..., f_{mn}\_R\_c_{mn})^1$, whose instance is determined as follows. Each tuple $b \in B$ contributes to an output tuple x, such that:
- $x[A] = b[A]$, for every attribute $A \in \boldsymbol{B}$
- $x[f_{ij}\_R\_c_{ij}] = f_{ij}\{\{t[c_{ij}] \mid t \in RNG(b,R,\theta_i)\}\}$, for every attribute $f_{ij}\_R\_c_{ij}$ of x.
∎

B is called the *base-values relation* and R is called the *detail relation*. Note that the MD-join is simply a GMD-join with $m=1$. GMD-joins can be evaluated using the following algorithm:

Algorithm 1. GMD-Join Evaluation Algorithm

```
initialize result relation X (cf. Section 3.1)
for all tuples r in R do
  for all rows x of X do
    for each θ  in (θ ,...,θ ) do
           i        1     m
      if θ (x,r) is true, update the aggregates
          i
          (f  ,..., f  ) of x in the MD-join.
            i1       in
    od
  od
od
```

With appropriate optimizations (indexing, etc.), this algorithm is very efficient. For large base-value tables, the algorithm should partition B and compute the result of each partition individually, thus permitting efficient in-memory computation at the cost of a well-defined increase in the scans of R [3, 4].

---

[1] Attributes are appropriately renamed if there are any duplicate names generated this way.

Proposition 2.1
If every tuple in B is distinct, the MD-join $MD(B(B), R(R), ((f_1(R.c_1), ...,f_n(R.c_n))), (\theta))$
corresponds to the following relational algebra expression (join represented by a selection and a Cartesian product):

$$[_BF[f_1(R.c_1), ...,f_n(R.c_n)] \; \sigma[\theta](B \times R) \cup ((B - \pi[B]\sigma[\theta](B \times R)) \; X \; N_I)] \tag{1}$$

where F is the aggregation operator (see [6]), and N_I is a one tuple n-ary relation
whose attributes are the initial values for the aggregates in l.
∎

An GMD-join can be composed with other relational algebra operators (and other
GMD-joins) to create complex GMD-join expressions. While arbitrary expressions
are possible, it is often the case that the result of an GMD-join expression serves as
the base-values relation for another GMD-join. This is because the result of the GMD-
join has exactly as many tuples as the base-values relation B. In the rest of this paper,
when we refer to (complex) MD-join expressions, we mean only expressions where
the result of an (inner) GMD-join is used as a base-values relation for an (outer)
GMD-join.

Next we generalize two effective algebraic optimizations for MD-joins proposed in
[4]:

Theorem 2.1 (Commutation of GMD-joins and Selections)
Let $B(B)$ and $R(R)$ be relations, θ_i be a condition with $attr(\theta_i) \subseteq B \cup R$, C be a condition with $attr(C) \subseteq R$, and l_i a list of aggregate functions. Then:

$$MD(B, R, (l_1, ...,l_m), (\theta_1 \wedge C, ...,\theta_m \wedge C)) = \tag{2}$$

$$MD(B, \sigma[C](R), (l_1, ...,l_m), (\theta_1, ...,\theta_m))$$

∎

This rule implies that the GMD-join can be computed using an indexed rather than a
full scan of the detail relation.

Theorem 2.2 (Coalescing GMD-joins)
Let $B(B)$ and $R(R)$ be relations, θ_i and $\underline{\theta}_i$ be conditions such that $attr(\theta_i) \subseteq B \cup R$ and
$attr(\underline{\theta}_i) \subseteq B \cup R$, and l_i and $\underline{l}_i$ be lists of aggregate functions. Then

$$MD(B, MD(B, R, (l_1, ...,l_m), (\theta_1, ...,\theta_m)), (\underline{l}_1, ...,\underline{l}_m), (\underline{\theta}_1, ...,\underline{\theta}_m)) = \tag{3}$$

$$MD(B, R, (l_1...,l_m,\underline{l}_1...,\underline{l}_m), (\theta_1...,\theta_m,\underline{\theta}_1,, ...,\underline{\theta}_m))$$

∎

This optimization would be hard to detect without using GMD-joins since a standard relational algebra expression (cf. Proposition 2.1) would contain multiple self-joins and unions that would defeat most query optimizers. The optimization allows us to coalesce a series of nested GMD-joins into a single GMD-join, thereby reducing the number of scans of the base table in Algorithm 1.

We conclude this section with two examples from the telecommunications area that illustrate the use of the GMD-join. We consider the analysis of network performance data over a database containing IP flows. Each IP flow is a sequence of packets trans-ferred from a given source (identified by a SourceIP address and a SourceAS autono-mous system) to a given destination (DestIP and DestAS). All packets in a flow pass through a given router, which maintains summary statistics about the flow: StartTime identifies the time at which the first packet of the flow was encountered, EndTime identifies the time at which the last packet of the flow was encountered, NumBytes identifies the number of bytes in the flow. When a flow terminates in our simplified scenario, the router dumps out a tuple containing the above information about the entire flow and adds it to a relation with the following schema:

```
Flow(SourceIP, SourceAS, DestIP, DestAS, StartTime,
EndTime, NumBytes)
```

For simplicity, we assume that the finest time granularity is in seconds. Given a data warehouse containing the data described above, one might want to ask OLAP queries such as the following;

Example 2.1 (Pivoting)
For each source IP that has sent packets to the autonomous systems with the DestAS numbers *1, 54* and *110,* we wish to compute the average traffic to each of these loca-tions. This is cumbersome to express in standard SQL, but is easily modeled in MD-joins, as follows:

$$MD_3(MD_2(MD_1(B,Flow, (avg(NumBytes)), \theta_1), \; Flow, (avg(NumBytes)), \theta_2), \quad (4)$$
$$Flow, (avg(NumBytes)), \theta_3))$$

Where:
$B = \pi \, [distinct \; SourceIP] (Flow)$
θ_1 is "Flow.SourceIP = B.SourceIP & Flow.DestAS = 29",
θ_2 is "Flow.SourceIP = MD_1.SourceIP & Flow.DestAS = 54", and
θ_3 is "Flow.SourceIP = MD_2.SourceIP & Flow.DestAS = 110"..
∎

Note that the GMD-join expression returns all source IPs, regardless of whether or not they have any traffic to the specified destinations. This issue will be discussed below.

Example 2.2 (Correlated Aggregates)
A more complex OLAP query is to ask for the total number of flows, and the number of flows whose NumBytes value exceeds the average value of NumBytes, for each

combination of source and destination autonomous system for each SourceAS and DestAS pair. We express this query using GMD-joins, as:

$$MD_2(MD_1(B, Flow, (avg(NumBytes)), \theta_1), Flow, (count(*)), \theta_2) \qquad (5)$$

Where:

$B = \pi$ *[distinct SourceAS, DestAS] (Flow)*,

θ_1 is "Flow.SourceAS = B.SourceAS & Flow.DestAS = DestAS".

θ_2 is "Flow.SourceAS = MD_1.SourceAS & Flow.DestAS = DestAS &
        Flow.NumBytes >= avg_MD1".

avg_MD1 is the average computed in the inner MD-join.

∎

3 Computing the GMD-join

In this section, we describe two approaches for computing GMD-joins: a C++ implementation that interfaces with a commercial DBMS through a programming interface, and an optimizer that reduces GMD-join queries to SQL.

3.1 The GMD-join Query Engine

We implemented an MD-join query engine on top of a commercial DBMS. The query engine extracts tuples from the DBMS using a standard program interface. Simple algebraic operations that the DBMS can handle efficiently (projections, selections, and grouping of the base-values and detail relations) are pushed down to the DBMS. An advantage of this approach is that the query engine can be built to work on any DBMS by adjusting the interface to the DBMS.

The evaluation of GMD-joins can be split into two distinct parts:
1. Construction of the *base-result structure*, and
2. Computation of the aggregates.

Recall the definition of the GMD-join. We refer to the result of the GMD-join $X(X)$ with the schema $X = (B, f_{11}\_R\_c_{11}, ..., f_{1n}\_R\_c_{1n}, ..., f_{m1}\_R\_c_{m1}, ..., f_{mn}\_R\_c_{mn})$, as the *base-result structure*. This can be constructed by submitting the query equivalent of the base-values relation B to the DBMS and retrieving the resultant tuples. As each tuple of b is fetched, it is extended with default values for $f_{11}\_R\_c_{11}, ..., f_{1n}\_R\_c_{1n}, ..., f_{m1}\_R\_c_{m1}, ..., f_{mn}\_R\_c_{mn}$ and inserted into X. In order to speed up computation we index the base-result structure on those attributes of B used in the theta conditions. In the prototype used for our experiments, hash-indexing is used.

The GMD-join query engine then computes the aggregates applying the following algorithm:

Algorithm 2. Evaluation of the aggregates in the MD-join query engine

```
for all tuples r in R do
   for each θ_i in (θ_1, ..., θ_m) do
      fetch the rows x of X that fulfill θ_i(b,r)and
      update the aggregates (f_i1, ..., f_in) of b in x
   od
od
```

This modification of Algorithm 1 takes advantage of the indexes on the base-result structure. Instead of a query processing time of $|B|*|R|$ for each GMD-join, we achieve $k*|R|$, where k is the average number of output tuples that a tuple t in R contributes to. As the experiments in Section 5 will demonstrate, this algorithm achieves good performance even for complex OLAP queries and scales nicely.

3.2   Reducing GMD-joins to SQL

An alternative to computing the GMD-joins is to reduce the queries to SQL. Note that it is not obvious that a simple and efficient reduction exists at all; the succinct formulation of complex OLAP queries was one of the main motivations for the development of extensions like EMF-SQL and the MD-join operator. Consider, for instance, the relational algebra expression of Proposition 2.1 that translates GMD-joins using multiple joins and unions. Obviously, an SQL implementation of GMD-joins based on this expression would not only be quite complex, but also likely very inefficient. However, a much better mapping exists for GMD-joins using outer joins together with the SQL CASE statement. Such a translation better reflects the semantics and evaluation of GMD-joins.

Theorem 3.1 (Reducing GMD-joins to SQL)
Let $B(\boldsymbol{B})$ and $R(\boldsymbol{R})$ be relations, θ_i be a condition with attr(θ_i) $\subseteq \boldsymbol{B} \cup \boldsymbol{R}$, and l_i be a list of aggregate functions. Then the GMD-join:

$$MD(B, R, (l_1, l_2, ..., l_n), (\theta_1, \theta_2, ..., \theta_m)) \tag{6}$$

Can be computed using the following SQL expression:

```
select   B, l_1', l_2', ..., l_n', count(*)
from     B right outer join R
group by B
```

Here l_i' is a list of functions, such that for each aggregate function $f_i(A)$ in l_i, there is a corresponding aggregate function $f_i(case\ when\ (\theta_i\ and\ (A\ is\ not\ NULL))\ then\ A\ else\ init)$ in l_i'. $init$ is the initialization value of the aggregate: 0 for sum, and NULL for count, max and min.[2]

∎

---

[2] Average is computed using a sum and a count.

The SQL translation of the GMD-join mirrors Algorithm 1. The SQL expression evaluates each aggregate f_i of each tuple t in R using the condition *(θ_i and (A is not NULL))*[3]. If the condition returns true, then the attribute aggregate $f_i(A)$ is updated as usual, otherwise the aggregate is passed an initial value which will not change its result. To ensure that all rows are preserved after the aggregation, we add a count(*) to the query[4].

Note that, as in Proposition 2.1, we assume distinct tuples in B. If B is not distinct, we would have to eliminate the duplicates in the SQL expression for the outer-join, and subsequently join the aggregate result given by the SQL expression above on B. For simplicity, we will not consider the case of non-distinct base-values tables in this paper.

However, the reduction is not yet complete. Without any conditions, the right outer join reverts to a basic Cartesian product. In order to make the SQL query efficient, we have to push down the theta conditions from the case statement to the where clause of the query, as defined in Theorem 3.2. This allows SQL to perform an outer-join rather than a Cartesian product, which is crucial to getting an acceptable performance. We note that standard SQL optimizers are not able to perform this kind of optimization.

Theorem 3.2 (Pushing down theta conditions to the where clause)
Let $B(B)$ and $R(R)$ be tables, C_i is a conditions in the theta condition θ, $(f_1,...,f_n)$ a list of aggregate functions, and W is a condition in the where clause (W can be true, as in Theorem 3.1). Iff $C_k \subseteq \theta_1 \cap ... \cap \theta_n$ of the aggregate functions $f_1,...,f_n$ then:

```
select    B, f₁(case when ((Cₖ and C₁) and (A is not null))
              then A else init),...,
          fₙ(case when ((Cₖ and Cₙ) and (A is not NULL))
              then A else init), count(*)
from      B right outer join R
where     W
group by  B
```

Is equivalent to:

```
select    B, f₁(case when ((C₁) and (A is not NULL))
              then A else init),...,
          fₙ(case when ((Cₙ) and (A is not null))
          then A else init), count(*)
from      B right outer join R
where     Cₖ and W
group by  B
```

∎

---

[3] For simplicity, we assume that there are no NULL values in the detail relation.
[4] This is required because SQL will eliminate rows from the result if no aggregate results are computed, whereas MD-joins always return the entire result table.

Together, Theorem 3.1 and 3.2 provide a simple and effective transformation of GMD-joins into SQL.

We use the following simple function to encapsulate the case statement in the aggregation in Theorem 3.1 (ga abbreviates guarded aggregation).

```
create function ga
  (c boolean, a integer, i integer) return integer
is
begin
  case
  when (c and (a is not null))
  then return a
  else return i
end
```

The boolean c represents the theta condition, a is the attribute being computed, and i the initialization value.

It follows that nested GMD-joins can be translated into series of nested outer joins with aggregates using the case expression. Consider, for example, the SQL version of Equation 4 in Example 2.1:

Equation 4 using SQL

```
create view B as
 select   distinct SourceIP
 from     Flow

create view X as
 select   Base.SourceIP,
          sum(ga(true,NumBytes,0)) as sum1,
          count(ga(true,NumBytes,null)) as cnt1, count(*)
 from     Base right outer join Flow
 where    Base.SourceIP = Flow.SourceIP and
          Flow.DestAS=29
 group by Base.SourceIP

create view Y as
 select   X.SourceIP, X.sum1, X.cnt1,
          sum(ga(true,NumBytes,0)) as sum2,
          count(ga(true,NumBytes,null)) as cnt2,
          count(*)
 from     X right outer join Flow
 where    X.SourceIP = Flow.SourceIP and Flow.DestAS=54
 group by X.SourceIP, X.sum1, X.cnt1

 select   Y.SourceIP, Y.sum1, Y.cnt1, Y.sum2, Y.cnt2
          sum(ga(true,NumBytes,0)) as sum3,
          count(ga(true,NumBytes,null)) as cnt3,
          count(*)
 from     Y right outer join Flow
 where    Y.SourceIP = Flow.SourceIP and
          Flow.DestAS=110
 group by Y.SourceIP, Y.sum1, Y.cnt1, Y.sum2, Y.cnt2
```

The above examples provide a clear indication of the primary weaknesses of the SQL translation: a multiplicity of nested views that makes it hard for the DBMS to efficiently optimize the query, coupled with the use of case expressions and outer joins. However, as we shall show in Section 4, by exploiting the expressive power of GMD-joins and using its algebraic rules one can significantly improve the performance of the SQL queries.

4 Optimizations

In this section, we describe the key optimizations to getting efficient performance from the SQL reduction, and introduce a new GMD-join optimization. According to Theorem 2.1 one approach to optimize the GMD-join query is to restrict the detail relation. In some cases, it is also possible to restrict the base-values relation. Consider the following query:

Example 4.1
We want to compute the average traffic through the destination autonomous systems with the id 29, for each SourceIP that uses it. Expressed in MD-joins, this can be written as:

$$\sigma\,[\theta_S]\,(MD(B, Flow, (avg(NumBytes), count(NumBytes)), \theta_X)) \tag{7}$$

Where:
$B = \pi\,[distinct\ SourceIP]\,(Flow),$
θ_S is "Count(*) > 0", and
θ_X is "Flow.SourceIP = B.SourceIP & Flow.DestAS = 29".
∎

In order to optimize the query, we could restrict the detail table flow using the condition "Flow.DestAS = 29". However, consider θ_S and θ_X. If a tuple b of the base-values relation does not fulfill the condition θ_X, it will end up with a count(NumBytes) equal to 0. Thus, the MD-join above can be rewritten as follows:

$$MD(B, Flow, (avg(NumBytes)), \theta_X)) \tag{8}$$

Where:
$B = \pi\,[distinct\ SourceIP]\,\sigma[\theta_B]\,(Flow)$
θ_B is "Flow.DestAS = 29", and
θ_X is "Flow.SourceIP = B.SourceIP & Flow.DestAS = 29".

We summarize this optimization in two theorems:

Theorem 4.2 (Restricting the base-values relation on GMD-joins)
Let $B(B)$ and $R(R)$ be relations, π_B be a projection or group operation on B, $\theta_i = C_{i1} \wedge C_{i2}$ be a condition with attr$(C_{i2}) \subseteq R$, $B = \pi_B$ (R), and l_i be a list of aggregate functions. Provided that we do not wish to preserve the rows of B for which no results are computed then:

$$MD(\pi_B \ (R), R, (l_1, l_2, ..., l_n), ((C_{11} \wedge C_{12}), (C_{21} \wedge C_{22}), ..., (C_{n1} \wedge C_{n2})) \tag{9}$$

$$\Rightarrow MD(\pi_B \ (\sigma \ [C_{12}|| ... || \ C_{n2}] \ (B)), \ R, \ (l_1, l_2, ..., l_n), \ ((C_{11} \wedge C_{12}), ..., (C_{n1} \wedge C_{n2}))$$

∎

This optimization eliminates those tuples of B that are not required in the GMD-join, prior to computing the operator itself. Obviously, if a tuple does not fulfill the condition $\sigma \ [C_{12}|| \ C_{22}|| ... || \ C_{n2}]$, it does not contribute to the result of the GMD-join.

Theorem 4.3 (Restricting the base-values relation series of MD-joins)
Let $B(B)$ and $R(R)$ be relations, π_B a projection or group operation on B, $\theta_i = C_{i1} \wedge C_{i2}$ be a condition with attr$(C_{i2}) \subseteq R$, $B = \pi_B$ (R), and l_i be a list of aggregate functions. Provided that we do not wish to preserve the rows of B for which no results are computed then:

$$MD_2(\ (MD_1(\pi_B(B), R, l_1, (C_{11} \wedge C_{12}))), R, l_2, (C_{21} \wedge C_{22})) \tag{10}$$

$$\Rightarrow MD_2(MD_1(\pi_B(\sigma[C_{12}||C_{22}](B)), R, l_1, (C_{11} \wedge C_{12})), R, l_2, \ (C_{21} \wedge C_{22}))$$

∎

Theorems 4.2 and 4.3 show how the base-values relation can be restricted if we are prepared to eliminate those rows of B that do not have any aggregates computed for them in the final result. In terms of an SQL-like query language, these theorems allows us to use conditions in the having-clause to reduce the join and aggregation operations of the query. Of course, in the general case deciding whether or not a relational query would eliminate these rows in the final result is undecidable. However, one would typically expect to find such restrictions defined at a high-level by the user himself (e.g., like in the natural language definitions of the query in Example 2.1 and Example 4.1); a simple solution would be to include some keyword in the query language indicating whether or not the user wishes to preserve rows in B.

This optimization is particularly useful when computing queries where the base-values relation has the potential to be huge (e.g. a data cube or sub-cube) and for complex GMD-join expressions. In this case, restricting the base-values relation with any applicable theta conditions allows us to significantly reduce the size of the base-results structure during computation.

Note that both theorems result in disjunctive conditions on the base-values and detail tables. Many commercial DBMS can not efficiently optimize disjunctive conditions and are unable to use indexes to optimize the selection. Theorem 4.2 and 4.3 can be specialized if each of the conditions C_{i2} is on the form "$A=x_i$", where A is an expression on R and x_i is a constant. In this case, we use the selection condition $\sigma[A \text{ IN } (x_1, x_2, ..., x_n)]$ on B and R instead of the disjunction. Thus, Example 2.1 becomes:

$$MD(B, \sigma [\theta_S] (Flow), (avg(NumBytes)), \theta_X) \qquad (11)$$

Where:
$B = \pi [distinct\ SourceIP] \sigma[\theta_S] (Flow)$,
θ_X is "Flow.SourceIP = B.SourceIP", and
θ_S is "Flow.DestAS IN (29, 54, 110)".

This modification permits the DBMS to make full use of the indexes in the base data. This optimization is also used in the GMD-join query engine implementation, for efficient extraction of tuples from the DBMS.

The algebraic transformations defined for GMD-joins yield SQL expressions. Recall the transformation of a series of GMD-joins in Theorem 2.2 to a single GMD-join. In SQL, this corresponds to an unnesting (or flattening) of the query. The reduction of the number of joins typically results in an improvement of the query processing time.

Example 2.1 in SQL with coalesced GMD-joins

```
create view Base as
  select    distinct SourceIP
  from      Flow

create view X as
select Base.SourceIP,
       sum(ga(Flow.DestAS=29,Flow.NumBytes,0)) as sum1,
       count(ga(Flow.DestAS=29,Flow.NumBytes,null))as cnt1,
       sum(ga(Flow.DestAS=54,Flow.NumBytes,0)) as sum2,
       count(ga(Flow.DestAS=54,Flow.NumBytes,null))as cnt2,
       sum(ga(Flow.DestAS=110,Flow.NumBytes,0)) as sum3,
       count(ga(Flow.DestAS=110,Flow.NumBytes,null))as cnt3
from      Base right outer join Flow
where     Base.SourceIP = Flow.SourceIP
group by Base.SourceIP
```

This result emphasizes that the optimization of complex GMD-join expressions is non-trivial. For example, coalescing nested GMD-joins (Theorem 2.2) limits the possibilities to restrict the detail and base-value relations (Theorem 2.1 and 3.2). As a result the outer-join may degenerate to a Cartesian product. Needless to say, it will usually be preferable to perform several right outer joins on the detail table rather than performing a Cartesian product. Note that this is not a problem for the native GMD-join algorithm described in Section 3.1.

A related query optimization problem occurs when restricting the base-values and detail relations with the theta conditions in the presence of coalescing. In the presence of a large number of coalesced GMD-joins, the use of Theorems 2.1, 4.2, and 4.3 will not have any effect (due to the disjunctive theta conditions being so universal that all tuples pass) or in the worst case may even cause a degradation in query performance. As a result, a cost-based query optimizer should be used in order to determine when coalescing, and the other GMD-join optimizations should be used.

5   Comparison Studies and Experiments

In the following section, we describe a set of experiments carried out to compare the performance of the SQL reduction of GMD-joins evaluated directly on a commercial DBMS, with the GMD-join query engine built on top of the same DBMS, and to study the performance of the various optimizations in both SQL and on the GMD-join evaluation algorithm. We derived a number of test databases from the TPC(R) dbgen program with 300.000 tuples in the smallest database (approx. 45 Mb) and 1.2 million tuples (approx. 180 Mb) in the largest database and ran a number of test queries against this database on a SUN platform. For each of the experiments except Figure 1, the queries have been run against a 100Mb database containing 600.000 tuples and with 10.000 groups. We note that the database is indexed on each of the grouping attributes used in our queries, and that all of the queries used were formulated to make use of these indexes. The absence of such indexes would have significantly degraded the performance of the SQL implementation, whereas the GMD-join query engine would largely remain unaffected, as it constructs its indexes on the base-values table on the fly. In each GMD-join operator, we computed an average (sum and count). We ran each query using the GMD-join query engine and its corresponding SQL reduction to evaluate comparative performance. We compare only queries evaluated using the methods in Section 3 as, in the general case, SQL queries translated using Proposition 2.1 would be inefficient.

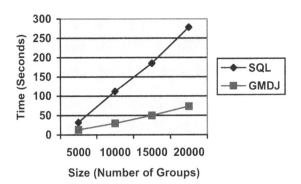

Fig. 1. Single GMD-join computation, comparing performance of SQL and GMD-joins when scaling up both the size of the database and the number of groups in the query

Figure 1 show the evaluation time for a simple MD-join query, when the size of the database is scaled up. The number of groups increases with the size of the database. As can be expected, the GMD-join query engine shows a linear increase in execution time. The comparatively worsening performance of the SQL reduction is explained by the increase in the size of the join being computed. Unlike SQL, the more specialized GMD-join algorithm does not perform a full join when computing the query.

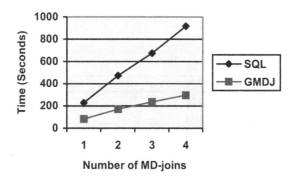

Fig. 2. Query processing time in seconds, with a dependent series of 1,2,3, and 4 GMD-joins

Figure 2 demonstrates the effect of an increasing number of GMD-joins in a query. The experiments were performed using queries containing a series of dependent GMD-joins (i.e., GMD-joins that cannot be coalesced). As above, the GMD-join query engine outperforms the SQL reduction; but both algorithms demonstrate a linear increase in processing time with the increase in the number of GMD-joins.

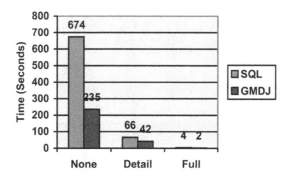

Fig. 3. Query processing times achieved for a query without applying restrictions, for the same query applying Theorem 2.1 (Detail), and for the query applying Theorems 2.1, 4.2, and 4.3 (Full).

Figure 3 shows the measurements on a complex OLAP query using 3 GMD-joins (without coalescing) when restricting the detail table and both the detail and base-values table. The selectivity of our conditions in the above experiments is approximately 2% over the source table. As can be noted, the use of these optimizations result in improvements and order of magnitude better than for the non-optimized query (this holds also for queries with larger selectivity as well). We find that the SQL implementation is able to make better use of these optimizations than the GMD-join query engine; probably because the restrictions allow us to significantly and efficiently reduce the size of the join.

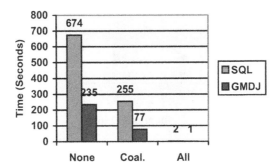

Fig. 4. Query processing times achieved for a query of 4 GMD-joins without restrictions and coalescing, the coalesced version of the query (Coal.), and the query processing time for the query with all the optimizations mentioned in this paper applied.

In our final experiment, we consider a series of 4 GMD-joins coalesced into a single GMD-join. As expected in this case, coalescing improves the query processing time by a factor of 4. By additionally restricting the base-values table and the detail table, we are able to significantly improve performance (in the query used in the experiment, we permitted restrictions similar to those in Equation 11).

6   Conclusion

Analyzing telecommunications network data often requires complex analysis of the very large data sets involved. We have previously observed that the GMD-join operator allows for tremendous flexibility in expressing OLAP queries such as those required in telecommunications applications, unifies a wide variety of OLAP queries in a single relational algebraic framework, and provides for efficient and optimized query plans when implemented using the algorithms given in [4].

In this paper, we present a reduction of GMD-joins to SQL, and show how GMD-join optimizations can be used to generate efficient SQL queries for a large range of GMD-join queries. As our comparison of GMD-joins with SQL show, an implementation of GMD-joins query execution is superior even to the most efficient reductions to SQL. We note, however, that making use of GMD-join optimizations in formulation of OLAP queries using SQL or query rewriting, can significantly improve the performance of such queries and achieve performance close to that of the specialized algorithm. Finally, we present an optimization on GMD-joins which improve query processing time by an order of magnitude, and discuss how this optimization interacts with the coalescing of GMD-joins. We note that using GMD-joins as the underlying algebraic framework permits significant and effective optimizations of OLAP queries.

We are currently working on a distributed query engine using the GMD-join query engine described in this paper, as well as further optimizations on the GMD-join query evaluation.

References

1. R. Caceres, N. Duffield, A. Feldmann, J. Friedmann, A. Greenberg, R. Greer, T. Johnson, C. Kalmanek, B. Krishnamurthy, D. Lavelle, P. Mishra, K.K. Ramakrishnan, J. Rexford, F. True, and J. van der Merwe, "Measurement and analysis of IP network usage and behavior", IEEE Communications Magazine, May 2000.
2. D. Chatziantoniou, "Ad hoc OLAP: Expression and evaluation.". In ICDE, 1999
3. D. Chatziantoniou, "Evaluation of Ad hoc OLAP: In-place Computation". In SSDBM, 1999
4. D. Chatziantoniou, M. Akinde, T. Johnson, and S. Kim, "The MD-join: an operator for Complex OLAP". ICDE 2001, 108–121
5. D. Chatziantoniou and K. A. Ross, "Querying multiple features of groups in relational databases". In VLDB 1996, 295—306
6. R. Elmasri and S. B. Navathe, "Fundamentals of Database Systems", Benjamin/Cummings Publishers, second edition, 1994
7. G. Graefe, U. Fayyad, and S. Chaudhuri, "On the efficient gathering of sufficient statistics for classification from large SQL databases". In KDD 1998, 204-208
8. J. Gray, S. Chaudhuri, A. Bosworth, A. Layman, D. Reichart, M. Venkatrao, F. Pellow, and H. Pirahesh, "Datacube : A relational aggregation operator generalizing group-by, cross-tab, and sub-totals". Data Mining and Knowledge Discovery, 1(1):29--53, 1997
9. T. Johnson and D. Chatziantoniou, "Extending complex ad hoc OLAP. In CIKM 1999
10. H. Wang and C. Zaniolo, "User Defined Aggregates in Object-Relational Systems". In ICDE 2000, 135-144
11. H. Wang and C. Zaniolo, "Using SQL to build new aggregates and extenders for object-relational systems". In VLDB 2000

Query Processing
in Embedded Control Programs

David Toman and Grant Weddell

Department of Computer Science
University of Waterloo, Canada
{david,gweddell}@uwaterloo.ca

Abstract. An embedded control program can be viewed as a small main-memory database system tailored to suit the needs of a particular application. For performance reasons, the program will usually define concrete low level data structures to encode the database, which in turn must be understood by anyone who needs to develop or modify the program. This is in contrast with the data independence that can be achieved by using a database system. However, because of space and performance requirements, the use of current database technology is not likely to be feasible in this setting. To explore one obstacle to this, we have developed a query optimizer that compiles queries on a conceptual schema to native Java or C code that navigates low level data structures. Of crucial significance is that an arbitrary collection of such structures, *perhaps already devised for an earlier version of the control program*, can be given as a part of the input to the optimizer. We present an overview of the underlying algorithms that are used to accomplish this. The algorithms are based on a novel resource bounded plan generation mechanism in which integrity constraints abstracting the definition of stored views are applied to source queries to extend the search space of possible query plans. We also report on some preliminary experimental results that suggest generated code navigates concrete data structures with an efficiency comparable to code written directly by expert programmers.

1 Introduction

Our contributions in this paper are twofold. First, we outline a new application area for database technology: the class of *embedded control programs* (ECPs). An ECP is a software subsystem consisting of one or more modules that implement a main memory database system tailored to suit the needs of a particular embedded control application. The main memory data is called *control data*.

An existing code base for an ECP will usually define low level data structures, a collection of C records containing arrays and pointers for example, to encode the control data. These structures must be understood by anyone who needs to develop or modify the ECP, which is in contrast with the data independence that can be achieved by using a database system. However, due to space and performance requirements, the use of existing database technology is not likely to be feasible in this setting. For example, there might be no (or very little)

W. Jonker (Ed.): Databases in Telecommunications II, LNCS 2209, pp. 68–87, 2001.

external storage available to the ECP. The main contribution of this paper is a new technique that enables one to compile high level declarative queries formulated on a conceptual view of an ECP's control data to a program in second or third generation programming languages that uses low level data structures to effect query evaluation. Our technique focuses on the issue of finding *navigational* query plans, those that utilize pointer navigation in places where standard relational query processing engines would need to scan a table or an index (a much more costly operation when accessing main memory data). An added complication relates to ECPs that are *legacy* systems with an existing and often very large base of code. In such circumstances, a query optimizer must be capable of *fine grained information integration*; that is, it must be possible to supply an arbitrary concrete encoding of main memory data already devised for an existing version of an ECP as part of the input to the optimizer.

To demonstrate that our technique is feasible and is capable of fined grained information integration, we have implemented a query optimizer for the **DEMO** system[1] that compiles SQL-like queries on a conceptual schema to native Java or C code that navigates an existing collection of low level data structures. This is in contrast with current ongoing efforts to develop "ultra-light" or main memory database engines in which data must be stored in product-specific internal forms. Our second contribution is a presentation of the underlying algorithms used by this optimizer. The algorithms are based on a novel resource bounded plan generation mechanism in which integrity constraints are applied to source queries to a *given limit* in order to extend the search space for possible query plans. An important role played by the constraints is to relate the "stored schema" that abstracts concrete data structures to a much more transparent conceptual schema on which user-defined queries are specified. Thus, the constraints are an abstraction of stored views with the additional property of being bi-directional; that is, they can be used to reason about conceptual data given view data and vise versa—a property that is crucial to our algorithms. As might be expected, the aforementioned limit on applying integrity constraints must be set very high to ensure completeness results (in cases where such limits exist). This remains true for simple families of queries. However, our experience indicates that low thresholds for this limit work well in practice.

The remainder of the paper is organized as follows. Section 2 gives a more thorough introduction to the topic of ECPs, and presents a pair of case studies that further clarify issues related to the processing of their queries. Our first case study also reports on some preliminary experimental results that suggest the code generated by our optimizer is capable of navigating legacy data structures with an efficiency that compares to legacy code written directly by expert programmers. In Sections 3 and 4, we outline the foundations of our approach to query optimization. In particular, we present the basic concept of resource

---

[1] **DEMO** is an acronym for *Design Environment for Main Memory Object Oriented databases*. The **DEMO** project at the University of Waterloo is an ongoing collaborative effort with Nortel Networks Ltd. [23] to adapt database technology for use in "industrial strength" ECPs.

bounded plan generation described above. Section 5 elaborates on this general approach with more details relating to the processor architecture, to plan selection and to code generation. Section 6 summarizes the contributions and gives a brief overview of related work.

2   Embedded Control Programs: Case Studies

To be more concise: an ECP is a software subsystem consisting of one or more modules that are responsible for managing a *uniform workload* for a *predefined set of transaction types* on a main memory database of control data. The assumption that there is a uniform workload is a key part of the definition that distinguishes an ECP from an arbitrary software subsystem. In particular, it becomes reasonable to ignore issues that relate to bulk reformatting of the control data. The following are examples of software systems that illustrate this. Each has a component ECP that consists of a single "dot h" file, the low level *schema* of the underlying control data, with a number of "dot c" clients that "implement" the predefined set of transaction types.

- An operating system kernel. The control data corresponds to information relating to open files, executing processes, and so on. The components of the ECP include the kernel functions and procedures together with utilities for managing the control data (a "ps" command, for example). In this case, the need to provide services for management and analysis purposes make it especially desirable to provide a conceptual interface to the kernel data. Other capabilities of a database system relating to concurrency control would also be useful.
- The control program for a communications system such as a telephone switch. In this case, the need for very high transaction throughput on so-called "office data" is clearly acute and therefore precludes using a database system that introduces *any* interpretive overhead on query evaluation. Virtually all the capabilities of a database system would be beneficial: physical data independence, non-procedural query languages, concurrency control and backup and recovery.

Both examples suggest that tools for processing queries in an ECP should perform as many as possible of the run-time tasks of existing database engines at the time the ECP is compiled or generated, particularly in the case of query optimization and plan generation. In the following subsections, we present a corresponding pair of case studies in order to clarify the issues related in particular to query processing in an ECP.

The first considers kernel data in the Linux operating system (the Linux kernel is the ECP for this case). We show how information on processes and open files may be viewed as a relational schema, and how queries on this schema are compiled by our optimizer to code that navigates the internal C records. We also report on some experimental results for this case that relate to a hypothetical re-implementation of the Linux ps(1) command as an SQL query on the relational

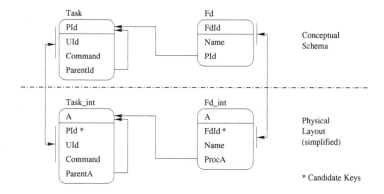

Fig. 1. Linux processes and file descriptors.

schema. The results indicate that the code generated by our query optimizer for this particular query navigates the Linux kernel data structures with an efficiency superior to the existing implementation. The second case study considers a more general class of ECPs: embedded control programs for *private branch exchanges* (PBXs). The case is a more dramatic means of clarifying issues related to query processing for ECPs, particularly the need to support the specification of legacy data structures as a parameter to query processing.

2.1 Case 1: The Linux Kernel

Our first case considers kernel data in the Linus operating system [3,4]. The kernel data structures contain hundreds of Kb of run-time data stored in highly compact data structures linked by pointers. However, a conceptual view of this data based on a collection of relation schemes is relatively straightforward to devise, which then enables the use of standard database application development techniques to be applied to the production, among other things, of Linux "system management" applications such as `ps(1)`, `top(1)`, `vmstat(1)`, and so on. In consequence:

– A clean and simple conceptual schema can be separated from implementation details, which in turn can be hidden from application developers to facilitate *data independence.*
– Application data retrieval requests can be expressed as high level queries over the conceptual view.
– The query optimizer, with the help of a small wrapper that attaches the generated code to the Linux kernel, provides a means for efficient execution of high level declarative queries over the actual kernel data structures.
– There is no interpretative overhead for query plan execution; the operations in our plans correspond directly to C statements.

For the remainder of this paper, we use examples derived from this case study; and for the sake of brevity, we focus on two data structures maintained by the kernel: *processes* running in the system, and *files* accessed by these processes. This simplified setting is depicted in Figure 1 in which the top part of the figure corresponds to the relevant part of a *conceptual schema* consisting of two relations:

```
Task(PId  integer, /* proc. Id */   Fd(FdId integer,    /* File Id */
    PPId integer, /* parent Id */     Name char(255),  /* file name */
    UId  integer, /* owner uid */     PId  integer)    /* owner PId */
    Cmd  char(16)) /* command */
```

Also, Linux ensures that every file descriptor has exactly one "owner" process, and that every process has a parent (the first process is its own parent). It is now possible to specify queries on this schema.

Example 1 A very simple example is a query that, given a user id, **uid**, prints out all tasks in the system owned by this particular user; a **ps(1)** replacement[2]:

```
select PId, ParentId, Command
from    Task
where   UId = :uid
```

Although the query resembles a simple selection from a base table, the generated code must perform a self-join of the **Task_int** tables to obtain the process id of the parent task. However, the query optimizer will recognize the opportunity to implement the self-join by navigating the **ParentA** field. Running the code generated by our query optimizer for this query against the actual data structures yields the following result (over a slightly modified Linux Kernel 2.2.5):

```
Enter uid: 501
  PId ParentId Command
  731      730 db2sysc
  732      731 db2sysc
  737      732 db2sysc
  736      732 db2sysc
 2336     2335 bash
```

It turns out that the running time is much shorter than the running time of **ps(1)**, although we suspect that this mainly due to avoiding the overhead associated with numerous system calls.

The relational abstraction of process and file descriptors, however, allows one to pose queries that would require a large amount of effort to express in C, but that remain easy to formulate in SQL:

1. For all processes, list the command that invoked their parent;
2. Find the "first" process in the system;
3. List all processes with no children; with one child; ... ,
4. List all files open by a particular process;
5. List all processes that share a particular file (based on file name); and
6. List pairs of processes that share an arbitrary file.

---

[2] For a particular choice of command line options.

In each case, our query optimizer will automatically generate C code that can be directly embedded in application programs without any additional runtime support. In addition, the query optimizer naturally favors navigational plans; the generated code often rivals code written by proficient C programmers, particularly in cases where complex navigational patterns need to be used.

The conceptual schema is connected to a corresponding physical schema using integrity constraints. The bottom part of Figure 1 illustrates this: the `Task_int` and `Fd_int` relations abstract the collections of data records used by the kernel, including explicit address fields (`A`, `ParentA`, and `ProcA`). The actual connection between the conceptual schema and the physical layout is more complex than suggested by Figure 1; we have simplified the way file names are obtained from a file descriptor (which in reality involves navigating several other data structures). Note that the *address* fields allow direct navigation between instances of the schemas in the lower part of the figure (cf. section 3.2).

Although our example in Figure 1 is limited, the approach appears general enough to capture all of the data structures used in the Linux kernel. We can provide relational abstractions for other kernel subsystems, e.g., for the memory management system, the IPC, the sockets and networking system, etc.

2.2   Case 2: A PBX

Now consider a more *data intensive* situation: a *private branch exchange* (PBX) for switching voice data at a private site. The software architecture for a typical PBX is illustrated in Figure 2. There are two subsystems that access a common

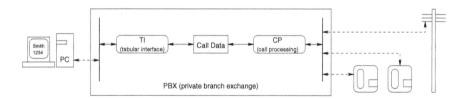

Fig. 2. Common Architecture for a PBX.

database of "call-data" storing diverse information about dial numbers, routing information, physical resources, test status information and subscribers. The first subsystem, *call processing* (CP), handles all clerical tasks relating to call setup, routing, hardware/software self diagnosis, and so on. CP handles transactions arriving from phone lines and handsets. The second subsystem, *tabular interface* (TI), provides external nested tabular views of various call-data and office parameters for network management purposes; subscriber data may be queried and new subscribers added by invoking transactions supported by this interface, for example.

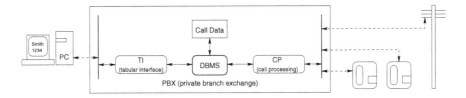

Fig. 3. A Preferred Architecture.

There are several problems with this architecture that arise from the typical situation in which the CP and TI subsystems are developed by programmers in a general purpose low level programming language such as C.

- Call data must be viewed by both systems in terms of its encoding as a set of C data (**struct**) types.
- The TI subsystem must be hand coded to provide the needed external views.
- Both subsystems must directly address any issues relating to concurrent access to call-data, and with its backup and recovery.

Now consider an alternative architecture illustrated in Figure 3. In this setting, a DBMS is used to manage the call-data. Although the introduction of a DBMS solves the above problems, the architecture of existing database technology results in an unacceptable reduction in transaction throughput for the PBX.

We would like to employ database technology in precisely the manner suggested by this alternative architecture, but with transaction throughput and store costs no less than what can be achieved with the basic architecture (in which expert programmers code in C). As we have noted earlier, this requires a great deal of what a database engine normally performs at runtime to be performed instead at compile-time. For example, access plans for statically specified queries must be "unfolded" directly into the TI and CP subsystems, and a simplified DBMS engine that excludes a query optimizer should be generated if there is no need to support dynamic ad-hoc queries. For the extreme case (static parameterized queries only, serialized transaction schedules suffice and no need for backup and recovery), there should be no DBMS footprint at all.

A further complicating factor, due to the infeasibility of complete rewrites of such systems, is that it must be possible to introduce this alternative architecture in an incremental fashion. Figure 4 illustrates an example of what we mean by this; only the TI subsystem for a hypothetical PBX software base has been reimplemented at the higher more abstract level supported by a DBMS. Thus, it becomes imperative to be able to relate a conceptual description of control data to an encoding corresponding to an existing collection of C types. The need to support very high transaction throughput also implies that a query optimizer is able to generate plans that navigate the C types with an expertise matching that of expert C programmers (existing systems are the performance benchmark).

There are other aspects of query processing in an ECP that derive from the embedded compile time nature of the application. The generated access plans

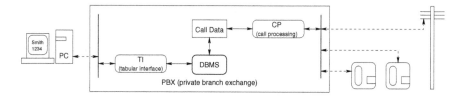

Fig. 4. An Incremental Architecture

should avoid the need to store arbitrarily sized intermediate results; that is, the temporary store required by a query plan should have a predictable constant size independent of the volume of control data. Conversely, although very high performance is needed for generated code, the amount of time spent on finding a query plan may be much longer than what would normally be acceptable in more traditional applications (e.g. in "interpreted" architectures in which queries are optimized at run-time).

3   Resource Bounded Plan Generation: Foundations

In this section, we focus on the basic ideas that underly our approach to finding valid (and preferably optimal) query plans for a particular storage layout. There are many alternative ways of describing how control data is encoded in an embedded controller's memory. Surprisingly, many of these definitions can be abstracted, without any loss of ability to find feasible query execution plans, by a combination of two basic ideas: the familiar notion of tuple and equality generating *generalized integrity constraints*, and notational conventions for capturing the *accessibility of relations*. We develop the foundations of our approach in this setting, beginning with the problem of finding query plans for conjunctive queries. We develop a query plan generator that performs a *resource-bounded expansion* of a query according to a supplied schema. We then outline extensions that are needed for discovering plans for general first-order queries. The following is used to demonstrate our approach for the remainder of the paper.

Example 2 (Running Example) Consider implementing a Linux application that lists files open by various processes in the system. The underlying conceptual schema is as follows:

```
Task(PId, UId, Command, ParentId)
Fd(FdId, Name, PId)
```

The schema captures the fact that, conceptually, the Linux kernel manages a set of *tasks* each associated with a (unique) process id (`PId`), a user id (`UId`), a parent task id (`ParentId`), and a command name (`Command`). Similarly, *descriptors of open files* are associated with an id, file name and the owning process id (cf. Figure 1). The following two relations

```
Task_int(A, PId, UId, Command, ParentA)
Fd_int(A, FdId, Name, ProcA)
```

capture the *physical layout* of the main-memory data: the values of the A fields represent *addresses* of records in main memory (cf. Section 3.2). Note that only the first two relation schemes are *visible* to the application developer; the later two are only visible to the query compiler.

We concentrate on the conjunctive fragment of first-order queries first. This fragment covers a wide class of *practical* queries, in particular those queries expressible using SQL's *select block*. It is well known [1] that conjunctive queries can be represented by a set of atomic formulas (representing their conjunction) and a set of *answer* variables.

Example 3 (Running Example: The Query) Our query in SQL, expressing the request for "what processes use what files?", is

```
select t.Command, f.Name
from   Task t, Fd f
where  t.PId = f.PId
```

This query can be represented by the pair $\langle (c, n), \{\mathsf{Task}(p, u, c, p'), \mathsf{Fd}(f, n, p)\}\rangle$ that in turn corresponds to a comprehension $\{((c, n) : \exists p, p', u, f.\mathsf{Task}(p, u, c, p') \wedge \mathsf{Fd}(f, n, p)\}$. There are two issues to clarify with this query.

 - The relations Task and Fd are not explicitly stored in memory; their contents must be reconstructed from their *internal* representation in the kernel memory (which, in general, may be much more complex than in this example).
 - The selection condition t.PId = f.PId has to be replaced by navigation of the ProcA field as the PId field is not stored in the Fd_int structures.

3.1   Conceptual Schema Declarations

To achieve our goals, the query compiler is provided with additional *schema information* that describes the relationship between the conceptual schema and the physically stored relations (cf. Definition 4 and Example 5), and the information about physically stored relations and the associated access methods (cf. Section 3.1). Even in cases where the underlying schema provides a direct access to every relation, this additional information will enable the query compiler to find alternative access plans (e.g., using navigation through other relations) which are often more efficient.

Integrity Constraints. The relationships between the conceptual schema and the actual encoding of the embedded application's data is described using generalized integrity constraints [1].

Definition 4 (Integrity Constraints) *Let P_1, ..., P_k, and P be atomic formulas (relation references) with variables as arguments. We use the following integrity constraints:*

$$P_1 \wedge \ldots \wedge P_k \Rightarrow R \qquad \text{(a tuple generating dependency)}$$
$$P_1 \wedge \ldots \wedge P_k \Rightarrow x = y \text{ (an equality generating dependency)}$$

The variables in constraints are implicitly universally quantified, except for the variables of R that do not occur in any of the P_i's, which are quantified existentially. We also require that variables x and y appear in one of the P_i's in the case of equality-generating constraints.

This definition covers a very large class of integrity constraints, including the usual *functional dependencies* and *embedded inclusion dependencies*. While we are aware of poor computational properties of such a large class of integrity constraints (including many undecidability results [1]), we take a more pragmatic approach based on a resource-bounded exploration of the consequences of the integrity constraints. Again, while theoretically infeasible, this approach seems to work well in practice.

Example 5 (Running Example: Integrity Constraints)   Here are some of the integrity constraints for our running example (cf. Figure 1):

```
Task(_,_,_,P) => Task(P,_,_,_)
Task(P,U,C,_) => Task_int(_,P,U,C,_)
Task_int(_,P,U,C,_) => Task(P,U,C,_)
Task_int(_,_,_,_,A) => Task_int(A,_,_,_,_)
Task(P,Q1,_,_), Task_int(_,P,_,_,A), Task_int(A,Q2,_,_,_) => Q1 = Q2
Task(P,Q,_,_), Task_int(_,P,_,_,A1), Task_int(A2,Q,_,_,_) => A1 = A2

Fd(I,N,_) => Fd_int(_,I,N,_)
Fd(_,_,P) => Task(P,_,_,_)
Fd_int(_,I,N,_) => Fd(I,N,_)
Fd_int(_,_,_,A) => Task_int(A,_,_,_,_)
Fd(F,_,P1), Fd_int(_,F,_,A), Task_int(A,P2,_,_,_,_) => P1 = P2
Fd(F,_,P), Fd_int(_,F,_,A1), Task_int(A2,P,_,_,_,_) => A1 = A2
```

We also assume that the first attribute of every relation is a key. Note that '_' denotes *anonymous variables* in the integrity constraints.

The constraints tell us, for example, that every file descriptor is associated with a single task, and that the task *pointer* in the internal representation of the file descriptor will indeed point to the correct internal task structure.

The Storage Model. To address the particular circumstance of main memory systems, one needs to specify how relations are physically stored in main memory. Recall that the goal is to be able to abstract virtually any data structure representing a collection of main memory *records* as a relation, and to support following "native C pointers" between such structures.

We therefore interpret every predicate in the plan as a collection of main memory record instances (e.g. instances of a C struct type) with the assumption that the first argument is the address of the record (and therefore a key) and that the remaining arguments are stored fields of the record; cf. Figure 1. While this requirement may seem very restrictive it is important to remember that other *conceptual* relations may be linked with these relations using integrity constraints. This can preserve a user's view of the conceptual schema of the embedded system in which physical addresses are not visible.

In particular, consider a predicate $P(x, y_1, \ldots, y_n)$ occurring in a query plan. The necessarily existing *binding pattern* for P (formally defined below) also implies the existence two "schema declarations:"

1. The functional dependency $(\{x\} \rightarrow \{y_1, \ldots, y_n\})$; and
2. The binding pattern $P^{bf\cdots f}$.

Therefore, in conjunction with appropriate inclusion dependencies, the addresses of records can be used to implement foreign-key joins. This fact may affect the way the conceptual schema and the integrity constraints for the main memory system are (or, more likely, should be) designed: we should make sure that we use the "navigational joins" for the most frequent foreign key joins performed in the system.

To define the *accessible* relations, one could give a set of relation names for which the embedded system has the ability to scan the instances of the corresponding relations. We generalize the notion of an accessible relation in the embedded system to take into account varying levels of *accessibility* of relations by using the familiar terminology for describing *binding patterns* in queries.

Definition 6 (Binding Patterns) *Let R be a relation of arity k and p a string in the alphabet $\{f, b\}$ of length k. Then R^p denotes a relation that limits access support to queries in which attributes corresponding to b positions in s are bound to a constant.*

Field Navigation. As pointed out above, one can expect that most accesses to stored tuples can be achieved by following pointers in the cases of key-foreign key constraints, rather than by executing iterators on the encodings of the base tables. This is one place where existing database technology often fails to match a skilled C programmer. However, the **DEMO** system is able to take advantage of the schema constraints and directly generate navigational code. There are two general cases that need to be considered.

- For the first occurrence of a variable in the plan that extracts a value from a corresponding field of a record, we simply generate an *assignment statement*.
- For any later occurrence of the same variable in the plan that corresponds to a value of another field extracted from another (possibly distinct) record, we generate an *if statement* that compares the value of the variable with the value in this field.

These two operations are justified by the $P^{bf\cdots f}$ binding pattern and the dependency ($\{x\} \rightarrow \{y_1, \ldots, y_n\}$). In addition, information about the cost of navigation vs. the cost of scanning a table is available to the query plan generator. The cost difference is usually so high that the plan generator picks navigation in almost all cases.

Iterators for Stored Relations. In addition, for any other binding pattern for P the embedded controller provides a *built-in* iterator that implements it.

Example 7 (Running Example: Binding Patterns)   We modify our example to take into account the existence of predefined binding patterns in our example. That is, we are given a set describing the allowed access to relations

$$B = \{ \text{Task\_int}^{fffff}, \text{Fd\_int}^{fffb} \}.$$

The first pattern corresponds to the fact that Linux provides an iterator that can walk through the tree of all running processes, the second to the fact that each process descriptor provides a pointer to an array of files open by this process.

3.2   Query Compilation

Integrity constraints are crucial to a fundamental part of the **DEMO** system: the module responsible for query plan generation. The module uses a chase-like [1,21] procedure that expands a query formula to an equivalent *conjunctive query* according to the schema integrity constraints. The goal is to identify sufficiently many *physically stored* relations in the expansion that enable a concrete means of determining the result of evaluating the original query. Even in cases for which the original query can be evaluated directly, this approach often leads to vastly superior query plans (recall that every attribute dereference is a significant overhead in main memory databases). We describe the general process in three steps, beginning with an introduction to the chase-like expansion procedure. We then show how the expanded query is used to find feasible plans, and also discuss families of constraints for which we can achieve completeness. (It is known that no complete procedure for finding execution plans for the general family of integrity constraints can exist.) At the end of this section, we extend our simple storage model to account for binding patterns of relations in the conceptual schema.

Definition 8 (Query Plan) *Let E be a query and B a set of binding patterns. We say that E satisfies B if E can be ordered in such a way that for each $R \in E$ there is a binding pattern $R^p \in B$ such that all attributes of R marked b appear in a preceding atom (in the order) or are parameter of the query.*

The plan generation algorithm takes as input a set of binding patterns. Note that there may be relations that do not appear in this set (i.e., don't have any allowed binding pattern, which is different from the $b \cdots b$ pattern).

Query Expansion. First we show that a conjunctive query can be *expanded* using the schema information to an equivalent conjunctive query that satisfies given binding patterns (i.e., represents a *query plan*). We use the schema constraints to define a set of rewriting rules on sets of atomic formulas as follows:

Definition 9 (Expansion Rules) *Let I be a set of integrity constraints and Q a set of atomic formulas. We define $\mathrm{Inf}(I)$ to be the set of rules*

$$Q \mapsto Q \cup \{R\theta\} \quad \text{for } (P_1 \wedge \ldots \wedge P_k \Rightarrow R) \in I, \text{ and } P_i\theta \in Q$$
$$Q \mapsto Q[x\theta \mapsto y\theta] \text{ for } (P_1 \wedge \ldots \wedge P_k \Rightarrow x = y) \in I, \text{ and } P_i\theta \in Q$$

induced by I; θ stands for renaming of variables in the integrity constraints to match those in the set Q; for embedded tuple generating constraints we require that variables, that are being substituted for the variable names in the constraint that only appear in the consequent, do not appear in Q.

We can expand every query Q using the above rules and form a set, $\mathrm{Exp}(Q, I)$, the closure of Q under $\mathrm{Inf}(I)$.

Unfortunately, in the presence of cyclic dependencies in the schema description (e.g., cyclic embedded inclusion dependencies), this set is not finite. Moreover, there is often no upper bound on the size of the expansion for the general constraint families; in Section 4 we discuss restricted families of integrity constraints for which such a bound can be found.

However, unlike many theoretical approaches, we do not immediately abandon the general case. Rather, we use a *resource-bounded approach* to control the depth of the expansion by restricting the number of applications of the $\mathrm{Inf}(I)$ rules. We achieve this by assigning a *rank* to variables in the generated conjunctive query as follows: rank 0 is assigned to variables in the original query, and then, each time a *new* variable is introduced in the expansion of an embedded tuple-generating dependency, the variable is assigned rank one higher than the maximal rank of variables that appear in a precondition of the instance of the applicable dependency. Moreover, if two variables are made equal using a equality-generating constraint, the higher rank is assigned to the result.

We define $\mathrm{Exp}^n(Q, I)$ to be the set of formulas $P(x_1, \ldots, x_k) \in \mathrm{Exp}(Q, I)$ such that $\mathrm{rank}(x_i) \leq n$. It is easy to see that the ranking of variables guarantees that the expansion procedure terminates after finitely many applications of the $\mathrm{Inf}(I)$ rules. The result of expansion, $\mathrm{Exp}^n(Q, I)$, is a set of atoms that represents a *conjunctive query* (defined as a conjunction of all elements of $\mathrm{Exp}^n(Q, I)$). To obtain the original query, it is also necessary to remove by projection all extra variables introduced during the expansion; this can be simply achieved by projecting on the free variables of the original query Q.

Plan Generation. We use the depth-bounded expansion procedure to find a valid query execution plan under the required restrictions of accessibility to relations in the conceptual schema. Our first *naive algorithm* shows that the problem can be essentially reduced to finding a correct join-order among the accessible relations equivalent to the original query. We can therefore reuse off-the-shelf join-order selection and cost estimation algorithms for this step.

Algorithm 10 (Naive Plan Generation)

FindPlan(Query Q, IC I, BP B, integer n) =
 $E := \mathrm{Exp}^n(Q, I)$;
 find $\min(c)$ and the corresponding plan P in
  $\{c : (P, c) = \mathrm{CostJoinOrder}(E')$
    where E' subsumes E, $E' \subseteq B$, and $Q \subseteq \mathrm{Exp}^n(E', I)\}$
 or fail if this set is empty;
 return(P);

where CostJoinOrder is a generic join-order selection algorithm.

In this algorithm, we exhaustively construct a plan for every subset of the candidate set E'. In practice, however, the selection of the candidate and the actual join-order selection can be interleaved, which makes the complexity of an exhaustive join-order selection dominate the combined complexity.

Lemma 11 (Soundness) *Let Q be a query, I a set of integrity constraints, B a set of accessible relations, and P a plan produced by* FindPlan(Q, I, B, n). *Then P implements Q on every database that satisfies I and B.*

When applied to our running example, the algorithm gives the following result:

Example 12 The expansion of the query $Q = \{\texttt{Task(P,\_,C,\_)}, \texttt{Fd(F,N,P)}\}$ from Example 2 with respect to the integrity constraints in Example 5 is

$$\mathrm{Exp}^1(Q, I) = \{\texttt{Task(P,\_,C,\_)}, \texttt{Fd(F,N,P)}, \texttt{Task\_int(A,P,\_,C,\_)}, \ldots\}$$

(the variables P, C, and N have rank 0, the rest of the variables have rank 1). Note the use of the equality generating constraint that matched the last attribute of Fd_int with the first attribute of Task_int. One of the sets E' that pass the conditions in Algorithm 10 is

$$E' = \{\texttt{Task\_int(A,P,\_,C,\_)}, \texttt{Fd\_int(\_,F,N,A)}\}.$$

This corresponds to the final (and indeed optimal) plan

$$\pi_{\{N,C\}}\texttt{Fd\_int(\_,F,N,A)} \bowtie \texttt{Task\_int(A,P,\_,C,\_)}$$

that utilizes the ability to navigate the address field in the Fd_int structure (cf. section 3.2).

In this setting, plan P_1 from Example 12 is no longer feasible, and the only feasible plan is P_3 with the join ordering

$$\pi_{\{N,C\}}\texttt{Task\_int(\_,P,\_,C,A)} \bowtie \texttt{Fd\_int(\_,F,N,A)}$$

that utilizes the ability to scan all file descriptors associated with a particular task by traversing an array of Fd_int structures associated with a given task (indicated by the binding pattern). Indeed, this is the situation in an unmodified Linux kernel.

In the actual system, we use a simple greedy plan generation algorithm that can be abstracted by the following pseudo code:

Algorithm 13 (Basic Greedy Plan Generation in DEMO)

GreedyFindPlan(Query Q, IC I, BP B, integer n) =
$\quad E := \mathrm{Exp}^n(Q, I)$;
$\quad E' := \emptyset$;
$\quad$ do
$\qquad C := \mathrm{GetCandidates}(E - E', B)$;
$\qquad$ if $C = \emptyset$ fail;
$\qquad E' := E' \cup \{\mathrm{BestCandidate}(C)\}$;
$\quad$ until ($Q \subseteq \mathrm{Exp}^n(E', I)$)

Note that by replacing the greedy *next goal* selection by a frontier generation, the greedy algorithm is converted to a branch-and-bound algorithm with only minimal changes to finding the feasible plans and the termination condition.

The actual algorithm used in **DEMO** avoids unnecessary recomputation of the $Q \subseteq \mathrm{Exp}^n(E', I)$ condition for the growing set E' using an incremental approach based on an efficient graph representation of conjunctive queries.

Code Generation. We now outline the final step in processing a query in the **DEMO** system: the generation of a low level target language code that implements an iterator protocol for a given query.

The plan generation algorithms guarantee that the plan given to the code generator is *feasible*—that all relations mentioned in this query plan satisfy the binding patterns. Consequently, at least one binding pattern must exist for every relation symbol mentioned in the plan.

Example 14 In our running example, the binding patterns are as follows:
$$B = \{ \mathtt{Task\_int}^{ffffff}, \mathtt{Fd\_int}^{ffffb} \}.$$

These declarations lead to two iterators: the `Task_iter()` providing references to all tasks and `Fd_iter(A)` providing references to file descriptors for a particular task. We also consider the case in which an additional iterator `Fd_iter()` providing access to all files (independently of tasks) is available.

The system provides a standard library for common situations (e.g., array scans, linked list scans, various search trees, etc), that can often be used directly or with only minor customization.

Example 15 Now we can finish our example. The following pseudo-code is generated by the **DEMO** system[3]. First, assume that we are able to access all file descriptors directly:

```
1.  Query(C,N) =
2.      for f ∈ Fd_iter do
3.          return(f.ProcA->Command, f.Name);
4.      od;
```

---

[3] In reality, the resulting C code must be "inverted" to form a proper iterator that follows **DEMO** iterator protocol.

The code illustrates the use of *scanning a base table* (2) and *stored field navigation* (3). However, if the file descriptors must be accessed through the corresponding process, the generated code is as follows:

```
1.   Query(C,N) =
2.       for t ∈ Task_iter do
3.           for f ∈ Fd_iter(t) do
4.               return(t.Command, f.Name);
5.           od;
6.       od;
```

Additional features include the possibility of using indices (where available), short-circuited evaluation of selection conditions, etc.

4   Restricted Families of Integrity Constraints

While the plan generation algorithm works quite well in practice, it cannot guarantee to find a feasible execution plan, even if one exists. However, for several restricted families of integrity constraints, one can determine a resource limit $n \in N$ based on the structure and size of the set of integrity constraints describing the schema that is large enough for Algorithm 10 to always find a plan if one exists. Finding such families has been the target of intensive research both in the database and the theorem proving communities (cf. Section 6 for discussion of related work in this area). One of these families of constraints can be characterized as follows: All the *tuple-generating* constraints are inclusion dependencies that fall into one of the following categories:

1. Full inclusion dependencies (all variables in the consequent of a constraint also appear in one of the antecedents);
2. Unary inclusion dependencies (at most one variable in the consequent of a constraint appears in any of the antecedents); or
3. Acyclic inclusion dependencies (no embedded constraint depends on itself).

It is also necessary for the equality-generating constraints to be *functional dependencies* with non-empty left-hand sides. For this restricted families of integrity constraints, our approach is guaranteed to find a plan if one exists:

Lemma 16 (Completeness) *Let Q be a query. Then, for every set of restricted integrity constraints I and every set of binding patterns B, there is $n \in N$ such that exactly one of the following situations occurs:*

1. *FindPlan(Q, I, B, n) finds a plan for Q if there is a valid plan, or*
2. *FindPlan(Q, I, B, n) fails if there is no valid plan for Q.*

The value of n depends only on I and B.

The value of n is bounded by a polynomial in the size of I for the family of full and unary inclusion dependencies, and is exponential in the size of I if one allows acyclic (non-unary) inclusion dependencies [1].

Assuming that only single-column statistics for join order selection are used (this disables the possibility of inferring that a particular binary relation is a contraction), the plan produced by Algorithm 10 is also optimal. As with completeness, this result cannot be generalized to the case of unrestricted integrity constraints. However, in those cases Algorithm 10 finds the best plan for the given resource bound n.

5   System Architecture

The overall architecture of the query processing component of the **DEMO** system is presented in Figure 5. There are essentially four components involved in the translation of a query to either C or Java code.

- The query is first parsed and translated to a canonical form in an extended relational algebra. Various standard simplifications to the query are performed at this time.
- The *complex query plan generator* then translates the high level algebraic expression to a low level form that expresses the chosen query plan. During this translation, the generator searches for conjunctive query subexpressions whose plan generation is delegated to the *conjunctive query plan generator*.
- The conjunctive query plan generator is a concrete realization of algorithm 13 outlined above. Optimized access plans for conjunctive subquery fragments of the general query are obtained by this module using the schema information. We elaborate on how this all works and on the interaction between the two plan generators below.
- The *code generator* finally converts the resulting query plan to the target language (currently, C or Java). In particular, the code generated from a plan consists of a pair of functions that realize an iterator protocol.

5.1   Plans for Complex Queries

The algorithms introduced in Section 3.2 focus on optimizing conjunctive queries. However, in a realistic setting, an optimizer must also handle negations and disjunctions in queries.

For general (first-order) queries, however, there is no hope for completeness of any planning algorithm (e.g., an equivalent of Lemma 16) due to the undecidability of satisfaction for relational calculus (there may be a feasible *empty* plan, but we are not able to detect it). Therefore, as with other approaches, we assume that for valid queries we must be able to find a plan for every atomic subformula (base table), or at least for every conjunctive subquery. In addition, the system uses several heuristics that propagate information between the conjunctive fragments of the original query to improve the chances of finding an execution plan.

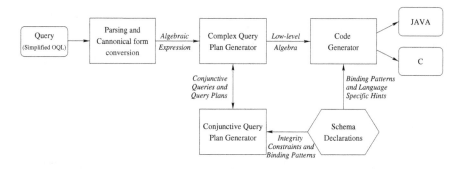

Fig. 5. DEMO Query Processing Subsystem Architecture.

The **DEMO** compiler uses a two-tier architecture to handle the general case. A complex query submitted to the *complex query plan generator* is processed inductively starting from the leaves and working towards the root as follows:

1. Find maximal conjunctive subqueries in the leaves of the given query;
2. Consult the *conjunctive query plan generator* and get plan(s) for these subqueries;
3. Generate subplans from these results for their immediate parents;
4. Replace the subqueries in the original query for which plans have been generated in Step 3 by views (that represent these plans); and
5. Repeat Steps 1 to 4 until the root of the original formula is reached.

In addition, information about variable bindings and other properties that hold in the answers to subqueries (and were determined by the plan generation procedures) are also recorded by the algorithms. This information is used to improve the plan in the upper levels of the query.

6   Summary Comments

In this paper, we have proposed a new application area for database systems, the class of *embedded control programs* (ECPs), and have presented an overview of how query processing is accomplished for such systems in the **DEMO** query optimizer. The focus of our discussion on query processing and our main technical contribution is the concept of *resource bounded plan generation*. Largely because of its ability to use binding patterns together with various families of tuple and equality generating dependencies, this new search paradigm has enabled us to deal effectively with two important problems:

1. The reverse engineering of existing legacy ECPs to the alternative architecture depicted in Figure 3; and
2. Competing effectively with expert programmers working directly in C.

In the first case, the variety of dependencies and binding patterns supported by our tools is sufficient to capture access to an arbitrary collection of C types, and to relate this kind of information to a conceptual schema. In the second case, we realized early on in the DEMO project that extensive semantic query optimization and "fine grained" plan algebras would be essential features of any query processing tools that could compete with expert programmers.

6.1 Related Work

Perhaps the most relevant ongoing work relates to various efforts to provide tools for information integration [2,8,10,13,14,15]. Of particular relevance is more recent work along this line that assume binding patterns limit how information sources can be accessed [7,17]. (See [20] for an excellent overview of this line of work, circa 1997.) Although the query semantics may differ (e.g. "find as much as you can" semantics), efforts along this line share the common goal of attempting to provide a uniform conceptual interface to existing legacy information sources. In particular, [6,16] approach the integration problem, including physical index selection, using similar techniques to those presented in this paper. Our work, however, differs in scale (e.g. we worry about accessing an array of C structs; others worry about accessing a file system or web document) and in the need, we think, to worry as much as we do about performance. In addition, the techniques described in this paper can be easily extended to handling duplicate semantics of queries [12,11] (which is beyond the scope of this paper).

Also of relevance is the extensive body of work on compiling logic programs [5,9,18,19,22]. Although a large focus of this work is on recursive query optimization (recursive queries do not seem particularly useful for ECPs) and there is only a limited effort to handle navigating legacy data structures, this work shares our own goal of effectively raising the level of programming to languages such as Datalog.

Acknowledgments. The authors gratefully acknowledge support by Natural Sciences and Engineering Research Council of Canada, by Communication and Information Technology of Ontario, and by Nortel Networks Ltd.

References

1. Serge Abiteboul, Richard Hull, and Victor Vianu. *Foundations of Databases.* Addison-Wesley, 1995.
2. Yigal Arens and Craig A. Knoblock. Sims: Retrieving and integrating information from multiple sources. In *SIGMOD*, pages 562-563, 1993.
3. Michael Beck, Harald Bohme, Mirko Dziadzka, Ulrich Kunitz, Robert Magnus, and Harold Bohme. *Linux Kernel Internals.* Addison-Wesley, 1997.
4. Remy Card, Eric Dumas, and Franck Mevel. *The Linux Kernel Book.* John Wiley and Sons, 1998.
5. Danette Chimenti, Ruben Gamboa, Ravi Krishnamurthy, Shamim A. Naqvi, Shalom Tsur, and Carlo Zaniolo. The ldl system prototype. *Transactions on Data and Knowledge Engineering*, 2(1), 1990.

6. Alin Deutsch, Lucian Popa, and Val Tannen. Physical data independence, con-
    straints, and optimization with universal plans. In *25th International Conference
    on Very Large Data Bases, VLDB'99*, pages 459-470, 1999.
7. Daniela Florescu, Alon Levy, Ioana Manolescu, and Dan Suciu. Query optimization
    in the presence of limited access patterns. In *SIGMOD Conference*, 1999.
8. Michael R. Genesereth, Arthur M. Keller, and Oliver M. Duschka. Infomaster: An
    information integration system. In *SIGMOD Conference*, pages 539-542, 1997.
9. Parke Godfrey, John Grant, Jarek Gryz, and Jack Minker. Integrity constraints:
    Semantics and applications. In Jan Chomicki and Gunter Saake, editors, *Logics
    for Databases and Information Systems*, chapter 9. Kluwer, 1998.
10. Laura M. Haas, Donald Kossmann, Edward L. Wimmers, and Jun Yang. Optimiz-
    ing queries across diverse data sources. In *International Conference on Very Large
    Data Bases*, pages 276-285, 1997.
11. Vitaliy L. Khizder, David Toman, and Grant Weddell. On Decidability and Com-
    plexity of Description Logics with Uniqueness Constraints. In *International Con-
    ference on Database Theory*, pages 54-67, 2000.
12. Vitaliy L. Khizder, David Toman, and Grant Weddell. Reasoning about Duplicate
    Elimination with Description Logic. In *Computational Logic 2000 (DOOD,'00)*,
    pages 1017-1032, 2000.
13. Alon Y. Levy, Alberto O. Mendelzon, Yehoshua Sagiv, and Divesh Srivastava.
    Answering queries using views. In *ACM Symposium on Principles of Database
    Systems*, pages 95-104, 1995.
14. Alon Y. Levy, Anand Rajaraman, and Joann J. Ordille. Querying heterogeneous
    information sources using source descriptions. In *International Conference on Very
    Large Data Bases*, pages 251-262, 1996.
15. Yannis Papakonstantinou, Ashish Gupta, and Laura M. Haas. Capabilities-based
    query rewriting in mediator systems. *Distributed and Parallel Databases*, 6(1):73-
    110, 1998.
16. Lucian Popa, Alin Deutsch, Arnaud Sahuguet, and Val Tannen. A chase too far? In
    ACM SIGMOD International Conference on Management of Data, pages 273-284,
    2000.
17. Anand Rajaraman, Yehoshua Sagiv, and Jeffrey D. Ullman. Answering queries us-
    ing templates with binding patterns. In *ACM Symposium on Principles of Database
    Systems*, pages 105-112, 1995.
18. Raghu Ramakrishnan, Divesh Srivastava, and S. Sudarshan. CORAL—control,
    relations and logic. In *International Conference on Very Large Data Bases*, pages
    238-250, 1992.
19. Konstantinos F. Sagonas, Terrance Swift, and David Scott Warren. XSB as an
    efficient deductive database engine. In *ACM SIGMOD International Conference
    on Management of Data*, pages 442-453, 1994.
20. Jeffrey D. Ullman. Information integration using logical views. In *International
    Conference on Database Theory*, pages 19-40, 1997.
21. Moshe Y. Vardi. *Fundamentals of Dependency Theory*, pages 171-224. Computer
    Science Press, 1987.
22. David H. D. Warren. An Abstract PROLOG Instruction Set. Technical Report
    309, Artificial Intelligence Center, Computer Science and Technology Division,
    SRI International, Menlo Park, CA, October 1983.
23. Grant Weddell. Main memory database management for embedded control appli-
    cations, 1998. CITO Research Project, Government of Ontario.

Benchmark for Real-Time Database Systems for Telecommunications

Jan Lindström and Tiina Niklander

University of Helsinki, Department of Computer Science
P.O. Box 26, FIN-00014 UNIVERSITY OF HELSINKI, Finland
{jan.lindstrom,tiina.niklander}@cs.helsinki.fi

Abstract. As long as there have been databases there has been a large interest in measuring their performance. Therefore, several different benchmarks have been proposed. However, previous proposals do not consider timely response and other telecommunication application requirements. This paper shortly reviews telecommunication requirements and previous work on real-time databases. These requirements are used to prepare the benchmark proposal for real-time database systems in telecommunication. The benchmark models a hypothetical telecommunication operator, which provides some services to subscribers. The services provided are selected to represent a wide range of real services from different application needs e.g. Intelligent Networks, 800 service and GSM user roaming.

1 Introduction

The requirements of the telecommunications database architectures originate in the following areas [8, 10]: real-time access to data, fault tolerance, distribution, object orientation, efficiency, flexibility, multiple interfaces, security and compatibility [1, 11, 15]. In summary, Network Services Databases is a soft/firm real-time system that contains rich data and transaction semantics, which can be exploited to design better methods for concurrency control, recovery, and scheduling. It has data with varying consistency criteria, recovery criteria, access patterns, and durability needs. This can potentially lead to development of various consistency and correctness criteria that will improve the performance and predictability of such systems.

The traditional database benchmarks, such as Wisconsin, TPC-A, TPC-B, and TPC-C (see [4]), do not consider the service time as correctness criteria. The response time or residence time is only used to describe the database performance. The response time is an important correctness criterion of a real-time database, where the client expects to receive the response within a given time limit.

Previous work on real-time databases in general has been based on simulation. However, several prototypes of general-purpose real-time databases have been introduced. The StarBase real-time database architecture [9] based on relational model has been developed over a real-time microkernel operating system. The performance of the StarBase database has been evaluated in [14]. RTSORAC [19] is a prototype of object-oriented real-time database architecture. It is based on open OODB architecture with real-time extensions and implemented over a thread-based

W. Jonker (Ed.): Databases in Telecommunications II, LNCS 2209, pp. 88-101, 2001.

POSIX-compliant operating system. Another object-oriented architecture called M^{2}RTSS [3] is a main-memory database system with classes that implement the core functionality of storage manager, real-time transaction scheduling, and recovery.

In this paper we present a benchmark that can be used to evaluate database performance for telecommunication services. The model covers most important features of the Intelligent Network concept and GSM roaming. These indicate the need for timely response and access to distributed data [17]. The services are modelled using simple transactions that represent the services. The benchmark also uses a workload model from telecommunication. In addition to the measurement of timely response, the database and service availability measurement is presented.

2 Telecommunication Requirements

The telecommunication field has different services, which have different database needs. The intelligent network (IN) concept models its services in a traditional fixed-line network. The GSM services are wireless. Telecommunication Management Network (TMN) has its own requirements for the databases. This section concentrates on the requirements for distributed databases. Requirements for separate databases are listed in [10, 12, 16].

The Intelligent Network (IN) [6] services do not necessarily require the support from a distributed database. A centralized database is enough to support them. The caller and called profiles can be fetched separately from their own databases at both ends of the call. Even the 800 Service does not require co-operation of multiple database nodes [7]. According to [10], the intelligent network has databases for traffic data, service data and customer data. The requirements analysis shows that transactions in the IN system are commonly small (only a few operations per transaction) and require short response time. This creates problems for most commercial database systems.

The best-known IN architecture is probably the Datacycle architecture by AT&T [2]. It is based on special hardware that allows fast data access. A regular type IN database has been presented by Norwegian Telecom Research [5, 18]. Their architecture is a shared-nothing parallel database where parallel relational database nodes communicate with each other via an ATM network. The traditional IN services can be implemented without transactions accessing multiple databases.

In wireless communication, like GSM or UMTS, roaming is an important basic service. Because the user may roam between different service providers, the service providers have a home location register for their own users and a visitor location register for the users currently in their network. A visiting user's move to a remote service provider requires an update on multiple places. These updates must be done in one atomic transaction to avoid losing the user's current location. The user is removed from the visitor location register at the departure service provider. It is added to the visitor location register at the arrival service provider. This change is also updated on the user's home location register. The authentication and roaming agreements are omitted from this paper, but a description can be found in [13]. The requirement analysis shows that mobile telecom transactions are commonly small and require short response time [12].

A Telecommunication Management Network may also need a distributed update operation, when a new link is established between two switches. At the minimum, this new link must be updated on routing information on the two participating switches. This update should be done atomically, to allow traffic passing over it. Of course, the update can be done sequentially but then some traffic going in one direction may not be able to return, until the update is performed on the other end also.

In a fixed network a conference call could need a distributed transaction to access the database on multiple service providers. Each participant of a conference call is using different operator. This requires an update to subscription information on the client databases of all operators used.

In televoting the distribution need can also exist. This scenario assumes that the vote collector is not a telecom operator. Each voter is allowed to vote only once. To reduce the load on the telecommunication network the vote collector and the telecom operator co-operate. A user registers his wish to vote on a telecom operator, which dispatches this wish to a vote collector. Then the user makes the vote, which is registered at the vote collector. The information that the user has voted is also registered to the operator. The operator can now prevent this user from voting several times without loading the vote collector and the communication lines.

Previous work has proposed that consistency of some subsets of the telecommunication transactions can be relaxed [10]. However, we recommend using full consistency on all transactions in the telecommunication database because it is easier to support the full consistency with all transactions than the full consistency with a subset of transactions and a relaxed consistency with another subset. Relaxed consistency would require additional semantic information of the transactions. This information can originate only from application developer. Application developer would need to include additional code to applications to maintain database consistency when integrity constraints are present. This would complicate and lengthen the development time. Therefore, the proposed benchmark requires full consistency with all transactions.

Compared to the traditional benchmarking of OLAP applications (TPC-B, TPC-C), which originate from banking applications and business applications, telecommunication databases contain much less data. This means that telecommunication databases fit nicely into the main memory of a computer or a cluster of computers [12]. Thus TPC-C and many other traditional benchmarks are focused too much on disk performance. Additionally, traditional benchmarks do not consider the service time as correctness criteria. Therefore, this paper proposes a new benchmark developed for telecommunication databases.

3 Benchmark

This benchmark models a hypothetical telecommunication operator. The network has multiple service providers. The service providers may belong to one operator or they may belong to multiple operators. Each service provider has its own database, but for this benchmark the databases are similar. The service provider has many customers, each with one or more subscriptions for different available services. The database represents the telecommunication services and billing information of each entity (service provider and service).

The components of the database are defined as consisting of five separate and individual classes: ServiceProvider, HomeProfile, VisitorProfile, ServiceInfo, and Subscriptions. The relationships among these classes are defined in the following UML-diagram (Figure 1). This diagram is a logical description of the classes. It has no implication to the actual, physical implementation.

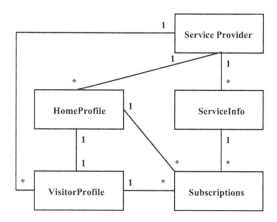

Fig. 1. Schema of the benchmark database.

Although the entity names in the model are collected from the wireless world, they can be used to model the IN services as well. The Virtual Private Network as described in [7] maps to our model by simply changing the names of the entities. The simple read and write transactions are also similar in both worlds. Likewise, the link establishment in TMN is only a distributed insert and occurs even less often than a user roaming in the wireless world.

In order for the transactions to represent a similar amount of work to all the systems, it is important that the records handled by the database servers, file systems, etc. are of the same size. Therefore, the records/rows must be stored in an uncompressed format.

The size of each item in the ServiceProvider class must be at least 100 bytes. The actual attributes of the class are not important for the benchmark itself. However, one of the attributes, called ProviderId, must uniquely identify the service provider and the information attached to it. In this benchmark we assume the ServiceProvider class (Table 1) to contain in any order or representation the attributes ProviderId, ProviderName and ProviderInfo.

Table 1. ServiceProvider class.

Data Attribute	Description
ProviderId	Unique identifier across the ServiceProviders
ProviderName	Name of the ServiceProvider
ProviderInfo	Additional information of the provider

The ServiceInfo class contains the information of the available service at each service provider. In this benchmark the services must be uniquely identified over all services on all service providers. The benchmark concentrates only on the service price and its usage in the service. The size of each item in the ServiceInfo class must be at least 100 bytes. For the benchmark it must have attributes (see Table 2): ServiceId, ServicePrice and ServiceName. The actual order of these attributes is not described.

Table 2. ServiceInfo class.

Data Attribute	Description
ServiceId	Unique identifier across the range of ServiceInfo
ServicePrice	Price of the service, at least 10 significant decimal digits and sign
ServiceName	Service name in uncompressed format

Each client of the system has one home service provider. The client information is located in the HomeProfile of that service provider. The size of the data item in the HomeProfile must be at least 100 bytes. The HomeProfile (see Table 3) contains the client's subscriber identity (SubsId), which identifies the client over all clients in the whole system. The clients also have a local identity (ClientId), which usually is much smaller than the SubsId. This id is used to connect the client with her local subscribed services. The home profile must also contain the client's true phone number, the current roaming position as the service provider id, and the client address as the connection information to the client.

Table 3. HomeProfile class.

Data Attribute	Description
SubsId	Unique identifier across the range of all HomeProfile
ClientId	Subscriber identification
PhoneNumber	Subscriber real phone number
CurPosition	Current position, i.e., provider identification
SubsAddress	Subscriber's address
SubscriberInfo	Additional information on the subscriber

The service provider keeps information about current roaming users in VisitorProfile (see Table 4). The class must at the minimum contain the identification of the roaming user and the identification of the user's own service provider. The size of each item in the VisitorProfile is assumed to be small, only 16 bytes. To map the visitors with the services, the visitors are also attached with a temporary ClientId for the mapping.

Table 4. VisitorProfile class.

Data Attribute	Description
SubsId	Visitor identification, the same as in HomeProfile
ClientId	Subscriber identification
HomeLocation	Service provider's ProviderId, used for locating HomeProfile

The class Subscriptions (see Table 5) connects the subscribers and services together. It must have the identification to the user and the service. These identifications together identify each data item in this class. The size of the data item must be at least 50 bytes. In addition to the identification attributes SubServiceId and SubClientId, the class contains information specific to this subscription. The information is stored in SubType, SubValue, and SubName attributes.

Table 5. Subscriptions class.

Data Attribute	Description
SubClientId	Identification of the subscriber from HomeProfile or VisitorProfile
SubServiceId	Identification of the service from Service class
SubType	Subscription type
SubValue	Connection information, normally real phone number
SubName	Subscription name

The data item identifiers of the ServiceProvider, ServiceInfo, Subscriptions, HomeProfile, and VisitorProfile must not directly represent the physical disk addresses of the items or any offsets thereof. The applications may not reference records using relative record numbers since they are simply offsets from the beginning of a file. For each nominal configuration, the test must use the minimum database size given in Table 6.

Table 6. Database size.

Table/Class	Number of rows
ServiceProvider	2
ServiceInfo	10
Subscriptions	50000
HomeProfile	30000
VisitorProfile	10000

The classes presented above are necessary in the databases of each service provider. The class descriptions, however, do not present the distribution aspect of the whole database system. The VisitorProfile of each service provider mainly points to HomeProfile and ServiceProviders outside the local database. The visitors are added to the VisitorProfile, when they enter the roaming area of one service provider. The current location is updated in the HomeProfile in the 'home' provider's database. On the service provider level there is also a connection, because roaming may not be allowed between two random service providers. The connections crossing over the database borders are presented in Figure 2.

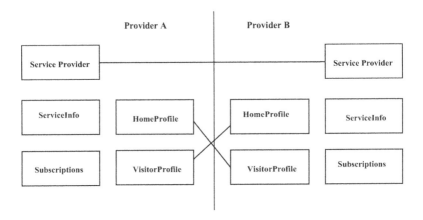

Fig. 2. Connections between two separate service provider databases.

The database contains a distributed integrity constraint between HomeProfile in the user's initial service provider and current location found from a remote service provider's VisitorProfile. Update to the current location in either class requires an atomic transaction to ensure the integrity of this constraint.

4 Test Transactions

The transactions represent the work performed when a customer uses some telecommunication services. The transactions are performed in a database of some service provider(s). This set of transactions presents the minimum that can be used for evaluating a distributed database for telecommunication. Each transaction has its own purpose in the test set.

The **GetSubscriber** transaction is used to search a specific subscriber from the database. It is mainly a simple local transaction, but can become distributed, if the query is given to a database that is not the home of the queried client. These transactions require subscriber identification, i.e. a Sid attribute as an input variable.

```
BEGIN TRANSACTION GetSubscriber
  SELECT PhoneNumber
   FROM HomeProfile
   WHERE SubsId = Sid;
  if NOT EXISTS then
   SELECT HomeLocation
     FROM VisitorProfile
     WHERE SubsId = Sid;
   SELECT PhoneNumber
     FROM HomeProfile@HomeLocation
     WHERE SubsId = Sid;
  end
END TRANSACTION
```

The **UpdateSubscriber** transaction is used to modify some of the subscriber data. This simple update transaction represents a large set of update transactions that access data in one class. The chosen update transaction updates the subscriber's name and number. It is actually the same as the *Modify Subscriber Number* service. This query also represents the *Customer Management* service in IN CS-1.

```
BEGIN TRANSACTION UpdateSubscriber
  SELECT SubscriberAddress, SubscriberAddInfo
    FROM HomeProfile
    WHERE SubsId = Sid;
    UPDATE HomeProfile
      SET SubscriberAddress = data1,
      SubscriberAddInfo = data2
    WHERE SubsId = Sid;
END TRANSACTION
```

The **GetAccessData** transaction can be used in three different scenarios. In the first scenario, the query is used to fetch the new destination number from the database in case the number given in the parameter is an abbreviated number (e.g. this query implements the *Abbreviated Dialing* service in IN). The query returns the destination number, which is active when the query is executed. The second scenario for the **GetAccessData** is to fetch the new destination number from the database in case the number given in the parameter is forwarded to another number (e.g. this query implements the *Call Forwarding* service). The query returns the destination number, which is active when the query is executed. Finally, the third possible scenario for the **GetAccessData** query is to fetch the new destination number from the database in case the number given in the parameter is a group number (e.g. this query implements the *Universal Access Number* service). When executed the query returns the current active destination number. All of these three scenarios in this database map to the same transaction. The actual phone number is fetched from the Subscription table.

```
BEGIN TRANSACTION GetAccessData
  SELECT SubValue
    FROM HomeProfile h, Subscription s
    WHERE h.SubsId = Sid and
          s.SubClientId = h.ClientId;
  if NOT EXISTS then
    SELECT SubValue
      FROM VisitorProfile h, Subscription s
      WHERE h.SubsId = Sid and
            s.SubClientId = h.ClientId;
  end
END TRANSACTION
```

The **RoamingUser** transaction is executed when a mobile user enters the area of another service provider. It is assumed that this transaction is executed at the home service provider. In the beginning of the roaming agreement the service providers have authenticated each other. This is done before this transaction is initialised. Firstly, the current location of the mobile user is determined. Secondly, the VisitorProfile data from the current visiting service provider's VisitorProfile is removed, i.e. this can be a distributed write. Thirdly, a new VisitorProfile data is inserted in the VisitorProfile of the service provider in the mobile user's new position. Fourthly, the current location in the HomeProfile is updated to the new location.

```
BEGIN TRANSACTION RoamingUser
  SELECT CurPosition
    FROM HomeProfile
    WHERE SubsId = Sid;
  DELETE VisitorProfile@CurPosition
    WHERE SubsId = Sid;
  INSERT INTO VisitorProfile@NewPosition
    VALUES(data)
  UPDATE HomeProfile
    SET CurLocation = PosId
    WHERE SubsId = Sid;
END TRANSACTION
```

In order for the transactions to represent a similar amount of work to all the systems, it is important that the fractions of different transactions are specified. We use the traditional 80-20 fractions between read and update transactions. The read-only transactions GetSubscriber and GetAccessData are 80 % of all transactions. Table 7 shows the fraction of the different transactions in the test load. The transactions access both local and distributed data. The difference between the access ratios on different transactions is presented in Table 8. The update transactions use the same distribution on reads, as well. The GetSubscriber and GetAccessData can be made distributed only by knowing the internal data distribution and executing the transaction not on the database where the actual data is stored.

Table 7. Transaction mix.

Transaction name	Fraction of all transactions
GetSubscriber	60
UpdateSubscriber	5
GetAccessData	20
RoamingUser	15

Table 8. Distribution of local and remote operations.

Transaction name	Remote read	Remote update	Local read	Local update
GetSubscriber	5	0	95	0
UpdateSubscriber	0	0	0	100
GetAccessData	5	0	95	0
RoamingUser	0	80	0	20

5 Database Properties

The ACID (Atomicity, Consistency, Isolation, and Durability) properties of transaction processing systems must be supported by the system under test during the running of this benchmark. These tests are adopted from the TPC-B benchmark (http://www.tpc.org). They are intended to demonstrate that the ACID properties are supported by the system under test and enabled during the performance period.

All mechanisms needed to ensure full ACID properties must be enabled during both the measurement and test periods. For example, if the system under test relies on undo logs, then logging must be enabled even though no transactions are aborted during the measurement period. When this benchmark is implemented on a distributed system, tests must be performed to verify that home and remote transactions, including remote transactions that are processed on two nodes, satisfy the ACID properties.

In addition to the traditional database properties a telecommunication database often implements some mechanism to increase data availability. The availability increment needs to be measured, as well.

5.1 Atomicity Tests

The system under test must guarantee that transactions are atomic; the system will either perform all individual operations on the data, or will assure that no partially completed operations leave any effects on the data. This is tested by a telecommunication transaction for randomly selected subscription and by verifying that the appropriate data items have been changed in the HomeProfile, ServiceInfo, Subscription, and VisitorProfile classes. Additionally, perform a telecommunication transaction for randomly selected subscription, substituting an ABORT of the transaction for the COMMIT of the transaction. Verify that the appropriate records have not been changed in HomeProfile, ServiceInfo, Subscription, and VisitorProfile classes.

5.2 Consistency Tests

Consistency is the property of the application that requires any execution of the transaction to take the database from one consistent state to another. A consistent state for the telecommunication database is defined to exist when:

- For all branches, the price of the service in the Service class is the same.
- Subscriber information in the HomeProfile and VisitorProfile, if it exists, is the same in all branches.

To verify the consistency of HomeProfile and VisitorProfile perform the following. First, verify that the HomeProfile and VisitorProfile class is initially consistent by determining the data items in the classes and verify that each client in the HomeProfile has at most one entry in all instances of VisitorProfile classes on separate service providers. Second, verify that HomeProfile and VisitorProfile classes are consistent after applying transactions to the database by performing the following: Using the standard driving mechanism, submit a number of telecommunication transactions equal to at least 1000 transactions and note the number of transactions that are reported as committed. If the number of committed transactions is not equal to the number of submitted transactions, explain why. Re-verify the consistency of the HomeProfile and VisitorProfile classes of each provider.

5.3 Isolation Tests

Operations of concurrent transactions must yield results which are indistinguishable from the results which would be obtained by forcing each transaction to be serially executed to completion in some order. This property is commonly called serialisability. Sufficient conditions must be enabled at either the system or application level to ensure serialisability of transactions under any mix of arbitrary transactions. The system or application must have full serialisability enabled.

For conventional optimistic schemes, isolation should be tested as described below, where transactions 1 and 2 are versions of the standard telecommunication transaction. Systems that implement other isolation schemes may require different validation techniques. The isolation tests are the following:

- Start transaction 1 and start transaction 2 before transaction 1 is committed. Transaction 2 attempts to read the same subscription record as transaction 1 is updating. Verify that transaction 2 reads the old value of the subscription. Verify that subscription information reflects the results of transaction 1's update only.
- Start transaction 1 and start transaction 2 before transaction 1 is committed. Transaction 2 attempts to write the same subscription record as transaction 1 is updating. Verify that transaction 2 is aborted and subscription information reflects the results of transaction 1's update only.

5.4 Durability Tests

The tested system must guarantee the ability to preserve the effects of committed transactions and ensure database consistency after recovery from any one failure:

- Instantaneous interruption in processing, which requires a system reboot to recover.
- Failure of all or part of the memory.

A transaction is considered committed when the transaction manager component of the system has written the commit record(s) associated with the transaction to a durable medium. A durable medium is a data storage medium that is either:

- An inherently non-volatile medium, e.g., magnetic disk, magnetic tape, optical disk, etc., or
- A volatile medium with its own self-contained power supply that will retain and permit the transfer of data, before any data is lost, to an inherently non-volatile medium after the failure of external power. A configured Uninterruptible Power Supply (UPS) is not considered external power, if its price is included in the total system price.

The intent of these tests is to demonstrate that all successful transactions have in fact been committed in spite of any single failure. The system crash test and the loss of memory test must be performed with a full terminal load using a fully scaled database. At the moment the failure is induced, there must be multiple local and remote transactions in progress. Distributed configurations must have distributed transactions in progress. For each of the failure types perform the following steps:

- Start submitting telecommunication transactions. On the driver system record committed transactions in a "success" file.
- Cause a failure selected from the list above.

- Restart the system under test using normal recovery procedures and measure the time required for partial and/or full restore.
- Compare the contents of the "success" file to verify that every record in the "success" file has a corresponding record in the database.

5.5 Availability Tests

On telecommunication a database that provides no services in case of one failure, is often not enough. The telecommunication environment requires more tolerable service. A fault-tolerant server can provide at least a reduced amount of service even during earlier specified failure situations. On such a continuous, or highly available, server the induction of one failure does not force the system to full recovery. However, the durability tests are needed to ensure that the effects of committed transactions are persistent even if a part of the system fails.

The continuous database server that guarantees data persistence must not loose the data even when the system is unable to provide service to its client due to the maximum concurrent number of failures it tolerates. The database content must still contain the modifications of all committed transactions. This can be verified using the consistency test with the failure induction test.

The availability measurement of a continuous server is actually the service level degradation during the failure situation and the duration of the failure recovery period. The measurement of the recovery duration can be done as black-box by using the service level degradation as indication of the system status. This is very coarse and should be avoided. Using a white-box approach, the duration is measured with internal information of the system functionality. The period starts when one recoverable part of the server fails and ends when the part is again a fully functional and operational part of the system.

6 Computation of Ratings

The reported metric, transactions per second (tpsT), is the total number of successfully committed transactions divided by the elapsed time of the interval. A committed transaction is counted as successful only when it has both started and completed during the measurement interval and it has completed before its deadline.

A transaction success ratio (successT) is the total number of successfully committed transactions divided by the total number of transactions entered to a system under test during the measurement interval. The reported value transaction miss ratio (missT) is simply one minus successT. It contains all unsuccessfully ended transactions.

It must be demonstrated that the configuration, when tested at the reported tpsT and missT rates, has the property of stable throughput despite small changes in the number of concurrently active transactions.

The measurement period must:
- Begin after the system reaches a sustained "steady state";
- Be long enough to generate reproducible tpsT and missT results;

- Extend uninterrupted for at least 15 minutes;
- For systems that defer database writes to durable media, the recovery time from instantaneous interruptions must not be appreciably longer at the end of the measurement period than at the beginning of the measurement period.

The duration of the recovery period must be reported, if applicable. On continuous or highly available servers, the reported duration is the recovery period of one recoverable unit. If a highly available server tolerates only a fixed number of concurrent failures, then the server's recovery period must be reported, when more failures happen.

7 Conclusions

The databases are essential for the current telecommunication services and timely response from the database is essential for the client using the services. The traditional database benchmarks do not consider the service time as correctness criteria. However, timely response is an important correctness criterion of a real-time database system such as Network Services Database.

The telecommunication field has different services, which have different database needs. The intelligent network concept models its services in a traditional fixed-line network, GSM services are wireless, and TMN services have their own requirements for the database. These different requirements are modelled in the benchmark proposed in this paper.

The benchmark consists of a simple database schema modelling different telecommunication service requirements. Similarly, the transactions (or services) are selected so that they represent as large a set of real services as possible. The proposed benchmark requires that the database maintain all ACID properties.

Because the telecommunication environment requires more tolerable services, a fault-tolerant server can provide at least a reduced amount of service even during specified failure situations. Therefore, we have designed availability tests for the benchmark.

As a conclusion, the proposed benchmark very well represents telecommunication requirements and challenges. Therefore, this benchmark will be a very good instrument when comparing different commercial and research databases for telecommunication use.

References

[1] I. Ahn. Database issues in telecommunications network management. ACM SIGMOD Record, 23(2):37-43, June 1994.

[2] T. F. Bowen, G. Gopal, G. Herman, and W. Mansfield Jr. A scale database architecture for network services. IEEE Communications Magazine, 29(1):52-59, January 1991.

[3] S. Cha, B. Park, S. Lee, S. Song, J. Park, J. Lee, S. Park, D. Hur, and G. Kim. Object-oriented design of main-memory dbms for real-time applications. In 2nd Int. Workshop on Real-Time Computing Systems and Applications, pages 109-115, Tokyo, Japan, Oct. 1995.

[4] J. Gray, editor. The Benchmark handbook for database and transaction processing systems, second edition. Morgan Kaufmann, 1993.

[5] S.-O. Hvasshovd, S. E. Bratsberg, and Ø. Torbjørnsen. An ultra highly available dbms. In Proceedings of the VLDB 2000, p. 673.

[6] ITU. Q-Series Intelligent network recommendation overview. Recommendation Q.1200. ITU, International Telecommunications Union, Geneva, Switzerland, 1993.

[7] M. Jarke and M. Nicola. Telecommunication databases – applications and performance analysis. In Databases in Telecommunications, Lecture Notes in Computer Science, 1819, pages 1-15, Edinburgh, UK, Co-located with VLDB-99, 1999.

[8] R. Kerboul, J.-M. Pageot, and V. Robin. Database requirements for intelligent network: How to customize mechanisms to implement policies. In Proceedings of the 4th TINA Workshop, volume 2, pages 35-46, September 1993.

[9] Y.-K. Kim and S. H. Son. Developing a real-time database: The StarBase experience. In A. Bestavros, K. Lin, and S. Son, editors, Real-Time Database Systems: Issues and Applications, pages 305-324, Boston, Mass., 1997. Kluwer.

[10] B. Purimetla, R. M. Sivasankaran, K. Ramamritham, and J. A. Stankovic. Real-time databases: Issues and applications. In S. H. Son, editor, Advances in Real-Time Systems, pages 487-507. Prentice Hall, 1996.

[11] K. Raatikainen. Database access in intelligent networks. In Proceedings of the IFIP TC6 Workshop on Intelligent Networks, pages 163-183, Lappeenranta, Finland, 1994. Lappeenranta University of Technology.

[12] M. Ronström. Design and Modelling of a Parallel Data Server for Telecom Applications. PhD thesis, Ericsson Utveckling AB, 1997.

[13] M. Ronström. Database requirement analysis for a third generation mobile telecom system. In Databases in Telecommunications, Lecture Notes in Computer Science, 1819, pages 90-105, Edinburgh, UK, Co-located with VLDB-99, 1999.

[14] S. Shih, Y-K. Kim, and S. H. Son. Performance evaluation of a firm real-time database system. In 2nd International Workshop on Real-Time Computing Systems and Applications, pages 116-124, Tokyo, Japan, October 1995.

[15] J. Taina and K. Raatikainen. Experimental real-time object-oriented database architecture for intelligent networks. Engineering Intelligent Systems, 4(3):57-63, September 1996.

[16] J. Taina and K. Raatikainen. Database usage and requirements in intelligent networks. In D. Gaïti, editor, Intelligent Networks and Intelligence in Networks, pages 261-280, Paris, France, 1997. Chapman&Hall.

[17] J. Taina and K. Raatikainen. Requirements analysis of distribution in databases for telecommunications. In Databases in Telecommunications, Lecture Notes in Computer Science 1819, pages 74-89, Edinburgh, UK, Co-located with VLDB-99,1999.

[18] Ø. Torbjørnsen, S.-O. Hvasshovd, and Y.-K. Kim. Towards real-time performance in a scalable, continuously available telecom DBMS. In Proceedings of the First Int. Workshop on Real-Time Databases, pages 22-29. Morgan Kaufmann, 1996.

[19] V. Wolfe, L. DiPippo, J. Prichard, J. Peckham, and P. Fortier. The design of real-time extensions to the open object-oriented database system. Technical report TR-94-236, University of Rhode Island, Department of Computer Science and Statistics, February 1994.

Replication between Geographically Separated Clusters -
An Asynchronous Scalable Replication Mechanism for Very High Availability

Anders Björnerstedt, Helena Ketoja, Johan Sintorn*, Martin Sköld

Ericsson Research and Development (UAB) Sweden
* Independent Database Technology AB

Abstract. In telecommunication systems such as Home Location Registers (HLRs) and AAA-servers (Authentication, Authorization, and Accounting) requirements on availability, real-time, scalability, consistency and persistence (durability) of the data storage are important. A base for high availability, real-time, scalability, and consistency can be achieved by using a distributed real-time main memory database system – implemented on a local cluster of shared nothing processors. Even higher availability and improved persistence can be achieved through an additional level of redundancy, combined with geographical separation. Two or more clusters are separated geographically to protect against site failure or site unreachability, due to *any* reason, including externally caused disasters such as earthquakes, bombs or fires. A wide-area replication mechanism ensures that the database is always consistent and nearly always complete (up-to-date), at all sites. The persistency requirement on telecommunication systems is usually not as severe as, for example, banking systems. On the other hand, the availability and real-time requirements are usually very high, with milli-second response times and fail-over times of no more than a few seconds when a site fails.

The protocol chosen for replication between the separate sites/clusters can impact both availability and performance. If strict synchronous replication (2PC or 3PC) is imposed on all geographically replicated transactions, then clients will be forced to wait a considerable time on replies from geographically distant sites. A synchronous protocol can also have a tendency to propagate problems "upstream" from one site to others. Finally, if the replication protocol becomes a bottleneck then this will undermine the throughput and scalability of the local cluster.

This paper presents an *asynchronous replication* mechanism that preserves the availability, scalability, and consistency requirements while at the same time achieving acceptable level of persistency/completeness.

The paper also presents the Ericsson TelORB[1] platform including a distributed soft real-time main-memory database system. TelORB and the replication mechanism described here, is already in service in commercial HLRs and other products.

---

[1] TelORB is a registered trademark of Ericsson. For more information see
www.telorb.com

W. Jonker (Ed.): Databases in Telecommunications II, LNCS 2209, pp. 102-115, 2001.
© Springer-Verlag Berlin Heidelberg 2001

1 Introduction

In telecommunication systems such as Home Location Registers (HLRs) and AAA-servers (Authentication, Authorization, and Accounting) requirements on availability, real-time, scalability, consistency and persistence (durability) of the data storage are important. A base for high availability, real-time, scalability, and consistency can be achieved by using a distributed real-time main memory database system – implemented on a local cluster of shared nothing processors. Even higher availability and improved persistence can be achieved through an additional level of redundancy, combined with geographical separation. Two or more clusters are separated geographically to protect against site failure or site unreachability, due to *any* reason, including externally caused disasters such as earthquakes, bombs or fires.

A wide-area replication mechanism ensures that the database is always consistent and nearly always complete (up-to-date), at all sites. The persistency requirement on telecommunication systems is usually not as severe as, for example, banking systems. On the other hand, the availability and real-time requirements are usually very high, with milli-second response times and fail-over times of no more than a few seconds when a site fails.

The protocol chosen for replication between the separate sites/clusters can impact both availability and performance. If strict synchronous replication (2PC or 3PC) is imposed on all geographically replicated transactions, then clients will be forced to wait a considerable time on replies from geographically distant sites. A synchronous protocol can also have a tendency to propagate problems "upstream" from one site to others. Finally, if the replication protocol becomes a bottleneck then this will undermine the throughput and scalability of the local cluster.

This paper presents an *asynchronous replication* mechanism that preserves the availability, scalability, and consistency requirements while at the same time achieving acceptable level of persistency/completeness.
The replication mechanism is implemented in the TelORB system and is used commercially in products such as HLRs and SCPs.

1.1 TelORB

TelORB[1] is a distributed (cluster based) soft-real-time platform suitable for telecom server nodes. TelORB includes:
- a distributed soft-real-time OS.
- a distributed main-memory database.
- an optimized and reliable inter-process communication protocol.

---

[1] For an overview of TelORB see [TelORB99] or go to www.telorb.com.

- cluster management software that supervises the health of the cluster and handles reconfiguration and system upgrades on the fly.
- a CORBA compliant ORB.
- the HotSpot Java Virtual Machine.
- a file system.

TelORB implements cluster-level fault tolerance in *software*. Standard off-the-shelf hardware is used, such as Intel Pentium processors and 100Mb Ethernet switches. The TelORB cluster has no single point of hardware failure. Application development is supported for Java, C++ or C.

Both database data and application processes are configured and distributed over the cluster based on a unifying concept we call the *distribution unit*. Data and processes are declared to be instances of user defined types that are associated with logical *distribution unit types*. At system configuration a distribution unit type is associated with a *processor pool* that contains all or a subset of the available physical processors in the cluster. Individual instances of persistent data and processes are mapped at runtime to distribution units that are logical sub-divisions of the processor pool that the distribution unit type is associated with. Which distribution unit an instance is mapped to is decided by the system (with hints from the application) to achieve a best possible load sharing and redundancy replication over the pool. Which physical processor a distribution unit maps to is determined at runtime and by what processors are available in the processor pool. All distribution units have at least two replicas residing on different processors in the cluster.

In the event of processor failure, the system will automatically (in about 200 msec) mask out the failing processor from the configuration. Any affected distribution units will use the remaining replica(s) until the processor has recovered. Should the processor not recover within a few minutes, due to some serious hardware problem, the system self-heals by re-establishing redundant replicas for the affected distribution units, over the remaining processors.

The system can also be reconfigured dynamically to increase capacity by adding new processors. Once a new processor is enabled the system will redistribute distribution units to exploit the added processor. This is transparent to the application since the mapping to specific distribution units is the same, it is only the mapping to the individual processors in the processor pool that is changed since the pool now contains an additional processor.

All TelORB software (including the OS and the database system) can be upgraded on-line without any disturbances to application traffic (including database transactions). The adding/removal/reconfiguration of processors in a TelORB cluster can be compared to the hot addition and removal of disks in a RAID system.

TelORB clustering supports a heterogeneous hardware architecture. It is even possible to incrementally migrate a complete running cluster from one architecture to another, without interruption of service. The kernel is currently ported to the processor architectures of Intel Pentium and Sun Sparc.

1.2 The TelORB Database System

The TelORB database system is an integrated part of the high-availability functionality of a TelORB cluster. It is a high-performance, distributed main-memory database system with an Object-Oriented data model. The distribution is transparent to applications and distributed ACID transactions [GR93] are very efficient through the main memory implementation. For the distributed transaction commit within the cluster, TelORB uses an optimised synchronous two-phase commit protocol. Classic pessimistic two-phase locking is used, but locking of cluster internal replicas is lazy and optimistic. This ensures that all objects and all cluster internal replicas of objects are updated according to the ACID transactional properties. For real-time sensitive applications that do not require complete consistency, a non-blocking, non-transactional read is also available.

The complete database is periodically backed up to disk. The backup is based on a consistent checkpoint and runs as a low priority job that does not disturb applications. The database records are stored in write protected kernel memory. A similar approach is discussed in [SS91]. All updating database operations pass through system calls that verify parameters. Read access is simply direct memory access, once a database record has been locked by a transaction. Physical consistency checks are also provided through checksums of every database record and of backup files. Logical consistency checks can be provided through defining relational dependences between attributes in different classes and through user-defined triggers that can provide integrity constraint checking.

On-line schema upgrade is supported and is an integral part of the TelORB upgrade mechanism that makes it possible to upgrade the schema, migrate instances, and upgrade applications accessing the database according to the old or new schema. All these upgrade tasks are performed as one operator step and without interrupting applications.

Relative persistency is provided through replication of data between processors and periodic backups to disk. Optional logging of a possible subset of the data to disk is also supported, but at a cost of reduced availability and loss of real-time characteristics for that data. Database replication [GARCIA MOLINA79] between geographically redundant sites provides even better characteristics and is the focus of this paper.

2 Geographical Redundancy

TelORB achieves a very high level of availability at the cluster level. But there is still a non-zero probability of cluster failure due to:

- Physical catastrophe, earthquake, fire, etc.
- Failure of the wide area network, making the cluster unreachable
- Very serious software bugs causing multiple simultaneous processor failures[1] in a cluster.

To overcome such problems of availability and reliability, there is a need for an additional level of redundancy at the wide area network level.

The essence of geographical redundancy is to have two or more sites (TelORB clusters in this case) so that if one site fails, then the service provided by that site is taken over by the other site(s). We use the terms "primary" and "standby" when talking about the sending and receiving site respectively. This does not mean that a site necessarily has to be purely a sender or purely a receiver at any particular time. Such a configuration is, however, the simplest and we label it *passive hot standby*. In such a configuration there are two clusters. At any time one cluster is the primary and the other cluster is the standby. The standby is *hot* because it is kept up-to-date continuously by the asynchronous transaction replication protocol. It is ready to take-over as primary the instant (adjusted for latency) so is requested. It is *passive* because the hot standby does not handle any original traffic concurrently with the primary.

We will only mention one more configuration, as the focus of this paper is the replication mechanism. In a *partitioned load-shared* configuration, both sites handle original traffic and both sites act as standby for each other concurrently. The database is split into two independent partitions. One site is the preferred primary for one partition and preferred standby for the other and vice versa. In case one site fails, the other site handles both partitions until the failed site has recovered. This kind of solution is also sometimes called *active hot standby*.

The main technical problems facing the design of the replication mechanism is how to replicate the database from primary to standby so that:

- Fail-over can be achieved with guaranteed database consistency.
- Fail-over can be achieved as quickly as possible providing very high availability.
- Fail-over can be achieved with as little transaction loss as possible.

---

[1] Multiple simultaneous processor failures due to hardware faults are very rare, but are also covered by redundancy mechanisms discussed in this paper.

- The replication protocol never blocks or slows down the primary.

- The replication protocol has minimal overhead.

- The replication protocol scales-up with the sites. The minimum bandwidth of the WAN link of course has to be increased as the size of the sites increase.

- After site failure, recovery and catch-up (reconciliation) must be reliable and as fast as possible, to re-establish geographical redundancy.

- The sites should otherwise be as autonomous as possible, to minimize failure dependency.

There is a tradeoff between maximizing availability and minimizing transaction loss. In systems such as HLRs, the availability requirement is strong enough that the occasional loss of a limited number of transactions at fail-over is tolerated.

2.1 A Quick Walk-Through: Passive Hot Standby.

Figures 1(a) to 1(d) illustrate the major events relevant to the passive hot-standby solution. Two TelORB clusters are used to provide the telecom-service and give redundancy. From the point of view of the external network, there is just one node. At any given time, only one of the clusters is allowed to terminate original traffic. The traffic is routed to the current primary.

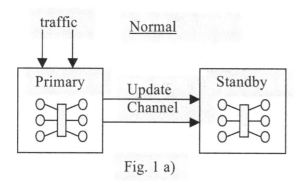

Fig. 1 a)

In figure 1(a) normal operation is illustrated. Traffic arrives at the primary and is handled by that site. For traffic that commit *updating* transactions, the update transaction is transferred to the standby. Complete and consistent transactions are transferred over a set of TCP/IP connections. The standby receives such transaction messages, which are checked for serialization [BHG87]. If the transaction serializes, it is applied as a "slave transaction" on the standby site. Transactions that do not immediately serialize at the standby are placed in a *backlog* queue. Normally all transactions serialize immediately and are applied immediately. Because transactions are not guaranteed to arrive at the standby in the same order that they committed on the primary, some transaction cannot immediately serialize and has to wait for some other transaction(s) that it depends on, to arrive and commit.

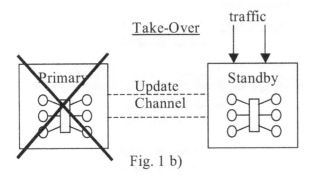

Fig. 1 b)

Figure 1(b) illustrates take-over. If the primary site crashes, becomes unreachable, or the operator (for other reasons) decides to perform a take-over, then the standby site takes over the duties of the primary. Traffic will be directed to the standby. If the primary did indeed crash, then of course there is no network redundancy until the primary has recovered. Once the old primary has recovered (which could be immediately after take-over if it was ordered manually) it contacts the standby and confirms that the standby has taken over.

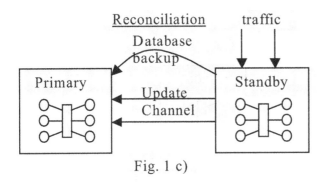

Fig. 1 c)

The two sites then start a catch-up procedure we call *reconciliation*, illustrated in figure 1(c). First the TCP/IP links are reconnected, but now from the site that was the standby (the new primary), to the site that was the primary (the new standby). Since transactions might have been lost at the old primary from the time it went down (if the take-over was not ordered manually) until the time it reconnected, there is a need for reconciliation. This is based on taking a backup at the old standby, and copying that backup to the old primary, and importing (from the backup), missing parts into the old primary.

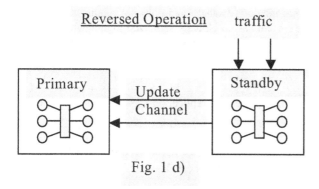

Fig. 1 d)

Once the backup has been applied and the backlog has disappeared at the old primary, then the old primary has caught up with the standby and is now acting as a passive hot standby to what used to be the standby, figure 1(d).

The situation is now reversed so that what used to be the standby is now the primary, and what used to be the primary is now the standby. The new standby is hot and ready to takeover. The system may continue this way indefinitely, or if the first primary is a *preferred* primary then a planned giveback may be performed at a suitable time. Such a giveback is logically the same thing as the takeover just described. A *planned* take-over or giveback can actually be done *smoothly* without losing any transactions. This avoids the reconciliation procedure and maintains uninterrupted redundancy.

2.2 Non-Blocking Wide-Area Replication Protocol

The protocol used by TelORB for replicating transactions *between* clusters is an asynchronous protocol, also referred to as lazy replication (as opposed to synchronous/eager). This means that a transaction at the primary does not delay and make its commit conditional on acknowledgement from the standby.

The local commit protocol at the primary is augmented as follows. After the critical phase, that is after the commit is irreversible at the primary, a replication message containing the complete transaction is forked off. The message identifies the read-set and includes all after-images. The replication message is sent without blocking and waiting for acknowledgement. Instead the transaction at the primary continues its normal termination and reports successful outcome back to the application. The commit at the primary is irreversible when the transaction has completed prepare for all locally involved processors and then secured at commit on at least two local processors. The replication message is sent after this, which ensures that replication messages always correspond to transactions that actually committed at the primary.

The TelORB database is a main memory database and does not require the transactions to log to disk. In fact the system has no globally serialized log anywhere and no global log sequence number. Thus there is no serialized log to use as basis for replication. Instead just parallel streams of commit messages are used, not guaranteed to arrive in order at the standby.

This non-blocking and un-serialized wide-area replication protocol has a number of advantages:

- It minimizes real-time interference with the application by keeping transaction commit time dependent only on the local cluster.
- It maximizes availability of data at the primary by allowing locks to be released as part of local commit.
- It is scalable, without any centralized bottleneck.
- Problems at the standby do not affect the primary.
- Communication problems between the primary and standby do not affect the primary.
- The protocol is simple. There is no transaction recovery part, which tends to be the most complex and problematic part.

This could be contrasted with a synchronous protocol, where the commit at the primary waits for acknowledgement from the standby. Problems in obtaining a lock at the standby would then delay or abort the primary transaction. A processor crash, or a full cluster crash, of the standby would require recovery activity at the primary. A synchronous protocol would then have much poorer performance and be much more complex. Its advantage would be to provide a higher degree of persistency. That is, it could guarantee that committed transactions are never lost, barring multiple cluster failure.

2.3 Scalability

Every processor in the cluster has a dedicated local process (PpClient) that collects and buffers the replication messages from processor-local application processes. This is possible because TelORB uses a data shipping implementation inside the cluster, where the transaction coordinator executes in the same thread as the application process and accumulates the complete transaction. At transaction commit it is the coordinator in the application process that generates the complete transaction replication message and sends this to the processor local PpClient. The PpClient sends batches of transaction messages over the TCP/IP connection to a peer process (PpServer) in the standby. It does this in a way that optimises the use of its TCP/IP connection. At the standby, the PpServer further routes each transaction message over the cluster to a set of processes at its disposal. The messages are routed based on information about locality of the data that is affected by a particular transaction.

When the cluster is scaled up, the application processes are automatically spread out over the additional processors. Each new processor starts its local PpClient that automatically requests the creation of a dedicated PpServer peer in the remote site and they establish their TCP/IP connection.

The replication protocol is scalable because adding a processor just adds a corresponding independent TCP/IP connection.

2.4 Autonomous Site Configuration.

The details of how one cluster is configured is unknown by the other cluster. A cluster is pre-configured only with a set of IP addresses to be used for *making contact* with the other cluster. The PpClient/PpServer TCP/IP connections are then established dynamically.

The configuration data that is shared between the clusters is kept to a minimum. It is in essence just the data that defines which persistent data is to be geographically replicated. The data to replicate are typically most application data plus the replication meta-data itself.

The lack of detailed configuration knowledge shared between the clusters is a feature because it reduces the risk that a configuration problem at one cluster causes problems at the other.

2.5 Consistency of the Standby

Due to the parallel independent TCP/IP links, transactions often arrive at the standby in a different order than they where committed and sent from the primary. Before being applied on the standby, each transaction has to pass a serialization check. This guarantees that mirrored transactions are applied in a serializable (isolated) order [GR93]. The serialization logic follows what amounts to an optimistic locking protocol, except transactions that currently fail serialization are not discarded, but are placed in a backlog buffer to be retried later. Only transactions that have a consistent read-set are cleared for application. Create and delete operations require some special handling that we have to leave outside the scope of the paper.

We know that every transaction will serialize sooner or later because the transaction did serialize on the primary. Once the transaction succeeds in serializing, it is applied using essentially the same cluster local transaction mechanisms as used on the primary to create the original transaction.

The serialization logic makes use of a TelORB proprietary non-blocking read primitive. This allows applications to see committed records without locking them (non-repeatable reads [GR93]). A large number of transactions may test for serialization concurrently without blocking each other or any other transactions. Because transactions that do not (yet) serialize are placed in backlog, the database at the standby is always consistent even if not always up-to-date.

2.6 Minimum Latency

If there is no backlog, which is the normal case, then the standby is lagging only by what amounts to the latency of the communication link between the clusters. This latency is normally far less than a second. If there is a hard take-over then normally no transactions are lost. TelORB also supports smooth takeover to allow controlled switch of cluster with guaranteed no loss of transactions.

As mentioned earlier, transactions can arrive in an order on the standby that causes a temporary backlog. When backlog does arise, the standby has sufficient spare capacity for catching up, as soon as the missing data that caused the backlog does arrive. Lasting backlog typically only appears when something out of the ordinary has happened on one of the clusters, such as a processor reload. In such cases, when transactions probably have been lost, the basic replication mechanism has to be complemented with a compensation mechanism.

2.7 Compensating the Incompleteness of the Basic Replication

Since the basic replication mechanism cannot by itself guarantee completeness, a second transport mechanism for *compensations* is used to guarantee an upper bound on lost transactions. The compensating mechanism based on backups that is used in reconciliation may also be used periodically during normal operation. The reason such compensating mechanisms are needed is that the primary mechanism, based on the asynchronous transfer of committed transaction messages, is not 100% reliable[1]. A processor crash at the primary or the standby, or an overload of a TCP/IP connection between the sites, could drop some transactions that have been committed on the primary, but have not yet reached or been applied on the standby. Such dropped or delayed transactions will be a rare event if the sites (and the link between them) have been dimensioned properly. But it *will* happen occasionally.

---

[1] This is an inevitable consequence of the protocol being non/blocking

Backups are taken periodically at each cluster. The process of regularly taking backups is important not only for network redundancy, but also to serve as a consistent recovery basis for the cluster. When a backup is taken at the primary, a check is simultaneously made of the standby to detect any data that is part of the checkpoint but missing from the standby. If a difference is detected and the difference is minor, then a compensating transaction (called a *quick sync*) is applied to the standby. If the detected difference is major, then the generated backup is copied to the standby and there merged into the standby database. This latter larger form of compensation we call an *external sync*. The merge done by an external sync follows similar serialization and idempotency rules as the primary mechanism uses to apply slave transactions.

Either the quick sync or the external sync will correct for missing updates that belonged to lost transactions. After this, the standby is guaranteed to be complete up to the checkpoint.

It is important to note that, even some transactions have been temporarily lost or delayed in the standby, the standby node is still always *transaction consistent* (except during on-going external sync) and ready for immediate takeover.

3 Related Work

The approach for achieving fault tolerance in TelORB is based on using off-the-shelf hardware with software implemented fault tolerance [HK93] and replication over at least two processors [LI95]. TelORB is a highly-available system and includes a distributed operating system, a database system, and a CORBA ORB. TelORB has been in service commercially for several years. A research system with similar features was presented in [LNPR99]. The database system is part of the support for providing very high availability as in [CLUSTRA00], but is also distributed using the same principles as for the operating system to provide a more homogeneous integration of application processes and their data. The database system in TelORB is a main-memory DBMS that scales over a shared-nothing cluster which gives better scalability than, for example [TIMESTEN00].

A *lazy replication* approach, with separate transactions at primary and standby, is used for geographical redundancy. This is also used by systems such as DB2 [IBM99] and Oracle [ORACLE98], but is different in the way transactions are re-serialized on the standby. The approach is also different than in Tandem [LYON90] since the protocol is asynchronous in order not to compromise the system availability. An *eager replication* protocol, with one atomic transaction over both primary and standby, would compromise the availability of the primary in case of failures in the communication network or of the standby node. In a partitioned database where inter-partition transactions are allowed, using only lazy replication will not be enough. This is discussed in [BKRSS99].

4 Conclusions and Future Work

The paper presented a mechanism for asynchronous replication between geographically separated clusters. The mechanism preserves the availability, scalability, and consistency requirements while at the same time achieving acceptable level of persistency/completeness. The paper also presented the TelORB system, which is a platform that includes a distributed soft-real-time OS and a distributed main-memory database. The mechanism is implemented for replication between geographically separated TelORB clusters. It is used in commercial systems.

Future work includes support for transactions over several database partitions and over multiple clusters (more than two). This will require extensions to the asynchronous replication with techniques for ensuring consistency between partitions. Further optimizations of the periodic compensations that complement the replication mechanism will also be done.

5 Acknowledgements

The TelORB system has been developed at Ericsson over a period of more than 10 years and involved hundreds of people. Here we acknowledge some persons whom have contributed significantly to the development of the database system part of TelORB. All of them are, or have been, employees of Ericsson unless otherwise indicated.

Noriko Akinaga, Johan Andersson (Independent), Hans Bjurström, Jan Ebenholtz, Per Emanuelsson (Softlab), Sverker Eriksson, Lars Hennert, Jan Isacsson (Independent), Peter Lind, Johan Linderoth, Ulf Markström, Håkan Matsson, Stefan Näslund, Per Rudahl (Independent), Jacques Salerian, Bo Samuelsson, Mats Strandberg, Mårten Sundquist, Kimmo Ulltjärn (Softlab), Jerker Wilander (Softlab).

6 References

[BHG87] Bernstein P., Hadzilacos V., and Goodman N.: *Concurrency Control and Recovery in Database Systems*, Addison-Wesley, 1987

[CLUSTRA00] Clustra Systems Inc., 620 3rd Street, Oakland, CA: *The Five9s Pilgrimage: Toward the Next Generation of High-Availability Systems*, Technical White Paper, 2000.

[BKRSS99] Breitbart Y., Komondoor R., Rastogi R., Seshadri S., and Silberschatz A.: Update Propagation Protocols For Replicated Databases, In *Proc. of ACM SIGMOD Int'l Conf. on Management of Data*, Philadelphia, Pennsylvania, June 1999.

[GARCIA MOLINA79] Garcia Molina H.: *Performance of Update Algorithms for Replicated Data in a Distributed Database*, TR STAN-CS-79-744, CS Dept., Stanford Univ., Stanford CA, June 1979.

[GR93] Gray J. and Reuter A.: *Transaction Processing: Concepts and Techniques*, Morgan Kaufman Publishers, San Mateo, CA, 1993.

[HK93] Huang Y. and Kintala C.: Software Implemented Fault Tolerance: Technologies and Experience, In *Proc. of the 23rd Int'l Symp. on Fault-Tolerant Computing*, Austin Texas, June 1993.

[IBM99] IBM, New Orchard Road, Armonk, NY 10504 (USA). *DB2: Replication Guide and Reference*, SC26-9642-00, June 1999.

[LI95] Lee I., Iyer R.: Software Dependability in the Tandem GUARDIAN System, In *IEEE Transactions on Software Engineering*, Vol. 21, No 5, May 1995.

[LNPR99] Lindström J., Niklander T., Porkka P., Raatikainen K.: A Distributed Real-Time Main-Memory Database for Telecommunication, In *Proc. of the Workshop on Databases in Telecommunications*, Edinburgh, September 6 1999.

[LYON90] Lyon J.: Tandem's Remote Data Facility, In *Proc. of IEEE Compcon*, 1990

[ORACLE98] Oracle Corporation, 500, Oracle Parkway, Redwood City, CA 94065. *Oracle8i^{tm} Advanced Replication*, Technical White Paper, November 1998.

[SS91] Sullivan M. and Stonebraker M: Using Write Protected Data Structures To Improve Software Fault Tolerance in Highly Available Database Management Systems. In the *17th Int'l Conf. On Very Large Databases (VLDB)*, September 1991

[TelORB99] Hennert L. and Larruy A: *TelORB – The distributed communications operating system*, Ericsson Review No. 03, 1999

[TIMESTEN00] TimesTen Performance Software: *Data Replication and TimesTen – High-Availability for the Next Generation*, Technical White Paper, 2000.

Yima[1]: Design and Evaluation of a Streaming Media System for Residential Broadband Services

Roger Zimmermann, Kun Fu, Cyrus Shahabi,
Didi Shu-Yuen Yao, and Hong Zhu

Integrated Media Systems Center and Computer Science Department
University of Southern California
Los Angeles, CA 90089-2561
[rzimmerm, kfu, cshahabi, didiyao, zhu]@usc.edu

Abstract. We describe and evaluate the implementation of a streaming media system called Yima, which consists of a scalable continuous media server and client components. We report on the real-life experiences that we gained from streaming near NTSC quality video and audio to residential locations within a metropolitan area. We investigated the feasibility of such streaming services with current broadband technology. We describe our experimental setup and the results, which indicate that streaming applications, such as video-on-demand, are not only technically feasible but also may be economically viable in the near future.

1 Introduction

Most currently deployed networks, such as the Internet and corporate Intranets, are based on IP. Furthermore, isochronous media types, such as video and audio, are becoming ubiquitous and need to be disseminated via these IP networks. Yima is a complete end-to-end system that addresses the issues of (a) storing and retrieving isochronous media types, (b) delivering such media types with their real-time constraints intact, and (c) rendering the media at the client location. Yima relys on commodity off-the-shelf (COTS) hardware components, such as standard personal computers, to provide a cost-effective solution. It is designed to support a wide range of media delivery bit rates from several hundred Kb/s (e.g., MPEG-4) to more than 20 Mb/s (e.g., HDTV streams compressed with MPEG-2).

Many of todays isochronous media streams are compressed at variable bit rates (VBR), such as — for example — specified by the MPEG-4 industry standard. Yima supports the streaming of VBR media via a simple yet flexible flow control paradigm. With this mechanism, the sender (server) need neither know about the compression scheme that is used for a specific stream, nor understand the details of a stream's file format. Therefore, new streamable media types, such as real-time haptic information [9], can easily be supported.

---

[1] In ancient Iranian religion, Yima is the first man, the progenitor of the human race, and son of the sun. This research has been funded in part by NSF grants EEC-9529152 (IMSC ERC) and ITR-0082826, and unrestricted cash/equipment gifts from IBM, Intel and SUN.

W. Jonker (Ed.): Databases in Telecommunications II, LNCS 2209, pp. 116–125, 2001.

This study presents an evaluation of Yima for video-on-demand type applications (e.g., distance learning, movie-on-demand) via currently available broadband connections to residential homes. There have been a number of reports of trials and deployments of video-on-demand services. However, little to no information is available about the architectural details and performance tradeoffs for these systems. For this study we have setup a client-server testbed that employs industry standards such as MPEG-4, ADSL (asynchronous digital subscriber line), RTP (real-time protocol [8]) and RTSP (real-time streaming protocol) over IP to evaluate the feasibility and performance of near NTSC quality video delivery to the home. Fig. 1 illustrates the experimental setup.

Section 2 introduces the Yima system architecture while Section 3 discusses the performance evaluation. Section 4 contains our conclusions.

2 System Architecture

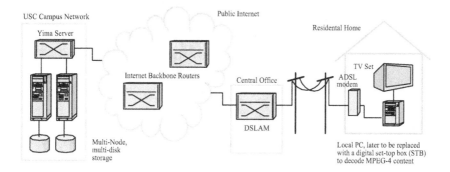

Fig. 1. Components of our experimental streaming media setup.

An important component of delivering isochronous multimedia over IP networks to end users and applications is the careful design of a multimedia storage server. The task of such a server is twofold: (1) to efficiently store the data and (2) to schedule the retrieval and delivery of the data precisely before it is transmitted over the network.

2.1 Data Placement and Scheduling

Magnetic disk drives have established themselves as the storage device of choice for continuous media (CM) servers because of their high performance and moderate cost. To achieve the high bandwidth and massive storage required for multi-user CM servers, disk drives are commonly combined into disk arrays to achieve many simultaneous I/O requests [3]. To efficiently store each individual multimedia object and to aggregate the bandwidth of multiple disks without requiring data replication, an object is commonly striped into n equi-sized blocks: $X_0, X_1, ..., X_{n-1}$ [6, 10]. Both, the display time of a block and its transfer time from the disk are a function of the display requirements of an object and the transfer rate of the disk, respectively. The display requirements of multimedia objects encoded with many of the popular compression algorithms, e.g., MPEG-4, vary as a function of time. Fig. 2 shows the display bandwidth requirement of a 10-minute segment for one of our test movies. The blocks of such variable bit rate (VBR) multimedia objects can be stored with two basic approaches: (1) the size of each block is varied to keep the display time per block constant, or (2) the size of each block is constant which results in a variable display time per block. The first approach simplifies the retrieval scheduling but the storage system becomes more complex. For the second approach the tradeoffs are reversed. The Yima framework is based on constant block sizes and our solution to VBR scheduling and delivery is presented in Section 2.3.

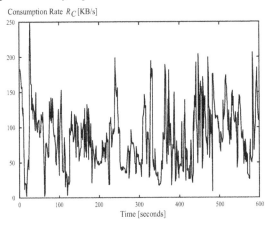

Fig. 2. Variable consumption rate of a 10-minute segment of a typical MPEG-4 movie.

There are two basic techniques to assign the data blocks to the magnetic disk drives that form the storage system: in a *round-robin* sequence [2], or in a *random* manner [7]. Traditionally, the round-robin placement utilizes a cycle-based approach to scheduling of resources to guarantee a continuous display, while the random placement utilizes a deadline-driven approach. In general, the round-robin/cycle-based approach provides high throughput with little wasted bandwidth for video objects that are retrieved sequentially (e.g., a feature length movie). Block retrievals can be scheduled in advance by employing optimized disk scheduling algorithms (such as *elevator*) during each cycle. Furthermore, the load imposed by a display is distributed evenly across all disks. However, the initial startup latency for an object might be large under heavy load because the disk on which the starting block of the object resides might be busy for several cycles. The random/deadline-driven approach, on the other hand, allows fewer optimizations to be applied, potentially resulting in more wasted bandwidth and less throughput. However, the startup latency is generally shorter, making it more suitable for interactive applications.

One disadvantage of random data placement is the need for a large amount of meta-data: the location of each block X_i needs to be stored and managed in a centralized repository (e.g., ⟨$node_x$, $disk_y$⟩). Yima avoids this overhead by utilizing a *pseudo-random* block placement. With random number generators, a seed is usually used to generate a sequence of random numbers. Such a sequence is pseudo-random because it can be reproduced if the same seed value is used. By placing blocks in a pseudo-random fashion, the next block in a sequence of blocks can always be found using the random number generator and the appropriate seed for that sequence. Hence, Yima needs to store only the seed for each file instead of a location for each block.

To achieve scalability, Yima servers are built as clusters of multiple server nodes. Each node may physically connect to one or more disk drives. A distributed file system provides a complete view of all the data on each node without the need to replicate individual data blocks (except as required for fault-tolerance [11]).

2.2 Real-Time Multimedia Delivery

In addition to the scheduling, data placement, and fault-resilience components, the Yima server software provides networking services to deliver data in real-time to end users. The stream data is transmitted via the industry standard networking protocols RTSP and RTP. These protocols are compatible with widely used industry standard technologies and hence provide compatibility and interoperability with other platforms in large-scale systems and the Internet, such as Apple Computer's QuickTime[TM].

RTP is used for the delivery of data packets because of the sensitive, real-time constraints of CM data. RTP uses the User Datagram Protocol (UDP) to ensure the quickest, although not necessarily the most reliable, packet delivery method. The UDP protocol features a low overhead but it delivers packets on a best-effort basis. Retransmitting packets is often considered inappropriate in the context of real-time streaming because it increases the latency between the sender and the receiver. This is especially critical for interactive and two-way applications. However, in a video-on-demand system some data buffering (and its effect of increasing the latency slightly) can often be tolerated if it improves the overall quality of the playback. In movie-on-demand systems the visual quality of the display is very important for the commercial success of such a service because end users can directly compare the playback with video displayed from a VCR or DVD player.

To study the effects of packet retransmissions we implemented two policies in our Yima server: (1) no retransmissions and (2) selective retransmissions based on a low-overhead protocol [5]. The results for both approaches are described in Section 3.

Hop #	Router
1	user-2iniv81.dialup.mindspring.com (165.121.125.1)
2	207.69.228.1 (207.69.228.1)
3	s4-1-1.lsanca1-cr1.bbnplanet.net (4.24.24.13)
4	p2-1.lsanca1-ba1.bbnplanet.net (4.24.4.5)
5	p0-0-0.lsanca1-cr3.bbnplanet.net (4.24.4.18)
6	s0.uscisi.bbnplanet.net (4.24.40.14)
7	usc-isi-atm.ln.net (130.152.128.2)
8	rtr43-18-gw.usc.edu (128.125.251.210)
9	rtr-gw-1.usc.edu (128.125.254.1)
10	zanjaan.usc.edu (128.125.163.158)

Table 1. End-to-end route from one Yima client (located in West Covina, connected via an ADSL line) to the Yima server (USC campus, Los Angeles). The distance between the two locations is approx. 40 km.

2.3 Client Buffer Management

The multimedia data that is streamed from the Yima server is consumed by an application at the client side. The client software implements the RTSP and RTP protocols to receive data transmissions from the server. A buffer module reassembles the RTP packets into multimedia blocks that are ready for consumption by the application.

Recall that the Yima multimedia delivery framework supports variable bit rate media. Numerous techniques have been proposed in the literature to transmit VBR media (for an overview see [1]). Most of the techniques *smooth* the consumption rate R_C by approximating it with a number of constant rate segments. Such smoothing algorithms need complete knowledge of R_C as a function of time to compute each segment length and its appropriate constant bit rate. In our implementation we reduce the smoothing technique into a binary variation: a stream is sent at a rate of either R_N or zero megabits per second. R_N is fixed at a value between the average and maximum consumption rate of a stream (the exact value depends on the client buffer size). A disadvantage of this technique is that it may temporarily require a higher link transmission rate than other smoothing techniques. Advantages include its simple design and that it can be applied dynamically to streams that are in progress without precomputation if, for example, only the maximum consumption rate is known.

Our technique is based on a simple flow control mechanism. A circular buffer of size B accumulates the media data and keeps track of two watermarks: buffer overflow WM_O and buffer underflow WM_U. Data are consumed from the same buffer by the decoding application. If the data in the buffer reach WM_O the data flow from the server is paused. The playback will continue to consume media data from buffer. When the buffer reaches the underflow watermark WM_U, the delivery of the stream is resumed from the server. If R_N is set correctly then the buffer will not underflow while the stream is resumed.

In an ideal case the server could be paused and resumed instantaneously and the watermarks could be set as follows: $WM_O = B$ and $WM_U = 0$. However, in a real world implementation the processing of pause and resume results in some delays (Yima uses the RTSP commands PAUSE and PLAY for these purposes). The accurate values of WM_O and WM_U are calculated as follows. After a pause request is sent out, the server is still sending data during a short time T_d. The data accumulated during T_d is $(R_N - R_C) \times T_d$. Hence, the buffer overflow watermark must be set to

$$WM_O \leq B - (R_N - R_C) \times T_d \qquad (1)$$

to avoid a buffer overrun. Similarly, the buffer underflow watermark must be set to

$$WM_U \geq R_C \times T_d \qquad (2)$$

Finally, the following condition must hold: $WM_O \geq WM_U$. If this condition is not met then the buffer size B is not large enough for this operating environment.

The delay time T_d depends on the network delay between the client and the server as well as the server request processing time and must be empirically obtained. In our experiments we measured $T_d \leq 2$ seconds.

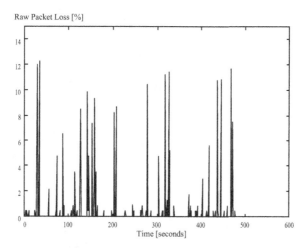

Fig. 3. Transmission (raw) packet loss for an MPEG-4 encoded movie segment of 10 minutes. The average is 0.365%.

3 Performance Evaluation

We have implemented a prototype of Yima to show (1) the flexibility of the Yima framework and (2) the feasibility of implementing the architectural design. First, we describe our experimental setup and then report the results.

3.1 Experimental Setup

Fig. 1 illustrates the details of our current system. For our Yima server setup, we are using two low-cost, commodity Pentium II 450MHz PCs with 384 MB of memory. Each PC is connected to an Ethernet switch (model Cabletron 6000) in our lab via a 100 Mb/s interface. Movies are striped over two 18 GB Seagate Cheetah disk drives (one per server node). The disks are attached through an Ultra2 low-voltage differential (LVD) SCSI connection that can provide 80 MB/s throughput. Red Hat Linux 6.0 is used as the operating system for each Yima server PC.

The data sent out from the Yima servers[2] in our lab are transported via the USC campus network to the public Internet. Table 1 shows the data route between one of our test client locations and the Yima servers. The geographical distance between the two end points is approximately 40 kilometers. The client was setup in a residential

---

[2] Note that the Yima architecture is scalable and we can easily expand the hardware to more PCs and/or disk drives.

apartment and linked to the Internet via an ADSL connection. The raw bandwidth achieved end-to-end between the Yima client and servers was approximately[3] 1 Mb/s.

For the client setup, we used low-cost, commodity Pentium III 600MHz PCs. The Yima player software runs on Linux or Windows and uses a variety of decoders to display different media types. Examples are MPEG-1 (1.5 Mb/s), MPEG-2 at DVD-quality (3-8 Mb/s) and at HDTV-quality (20 Mb/s), MPEG-4 (< 100 Kb/s to > 1 Mb/s), and Apple QuickTime[TM] formats (up to approximately 2 Mb/s). For our tests we chose an MPEG-4 software decoder called "DivX;-)" [4]. The high compression ratio of MPEG-4 allows for near NTSC quality video transmissions at less than 1 Mb/s bandwidth requirement, suitable for a residential ADSL connection. For example, our test stream was encoded with a frame size of 720 × 576 pixels and 25 frames per second (fps). The stream required an average of 105 KB/s (840 Kb/s) bandwidth for both the video and audio layers (audio was encoded in MPEG-1 Layer 3 format), shown in Fig. 2. This compares favorably with MPEG-1 which would require 1.5 Mb/s for a lower video resolution of 320 × 240 pixels at 30 fps. The buffer watermarks were calculated based on these parameters: $T_d = 2$ s, $R_C = 840$ Kb/s, $R_N = 1000$ Kb/s, and $B = 8$ MB. This resulted in $WM_O = 8152$ KB and $WM_U = 210$ KB.

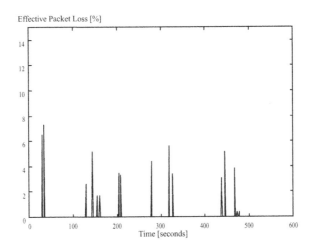

Fig. 4. Effective packet loss for an MPEG-4 encoded movie segment of 10 minutes. The average is 0.098%.

3.2 Results

We have conducted several end-to-end video streaming playback tests with the physical setup described in the previous paragraphs.

Since the data transfer is based on RTP/UDP/IP, a data packet could arrive out of order at the client or could not arrive at all since UDP does not guarantee packet

---

[3] The ADSL provider did not guarantee any minimum bandwidth but stated that 1.5 Mb/s will not be exceeded.

delivery. The characteristics of real-time data places rigid constraints on packet delivery and may be sensitive to packet loss. If a packet arrives after the decoder needs it, then it is considered a lost packet.

To achieve the best visual and aural quality possible at the client side we implemented the following two techniques:

a. The decoding of a movie was started with a delay after an initial amount of data had arrived at the client side buffer.
b. To reduce the number of RTP data packets lost between the server and client locations we implemented a selective retransmission protocol [5].

The initial buffer size was determined so that no buffer underflow starved the MPEG-4 decoder. The selective retransmission protocol was configured to attempt at most one retransmission of a lost RTP packet if it seemed likely (based on the round-trip packet delay) that the retransmitted packet would arrive in time for consumption by the decoder. With this technique the *raw* transmission data loss $P_0 = p$ can be reduced to an *effective* data loss of

$$P_1 = p \times (q + ((1-q) \times p)) \tag{3}$$

Where p is server-to-client and q is the client-to-server data loss probability. Eqn. 3 assumes a different loss rate for the two link directions since most broadband technologies (such as ADSL and cable modems) provide asymmetric throughput for upstream and downstream data flow. If the loss probability is identical in both link directions, then Eqn. 3 reduces to

$$P_1 = p^2 \times (2-p) \tag{4}$$

Figs. 3 and 4 illustrate the reduction in lost data from an average of 0.365% to 0.098% with this protocol in our test system. Note that the loss characteristic of a real network link is very bursty (as illustrated in Fig. 3) as compared with the uniform approximations of Eqns. 3 and 4. Hence, the real reduction of the loss rate is less dramatic than would be expected from the analytical equations. However, in practice the protocol worked very well and some of the loss peaks are completely eliminated while others are reduced significantly.

3.2.1 Visual Results

The visual and aural quality of an MPEG-4 encoded movie at less than 1 Mb/s is surprisingly good. Our test movie, encoded at almost full NTSC resolution, displayed some blocking artifacts during complex scenes that can be attributed to the high compression ratio. If a highly compressed movie is streamed over a network and packets are lost, then severe artifacts may be visible. In our setup, because of the low packet loss rate, almost no degradation of the movie was noticeable compared with its local playback.

4 Conclusions and Future Work

In this study we presented an evaluation of Yima which addresses the complete end-to-end issues of storing, retrieving, and delivering isochronous media types over IP networks. We described the architecture that is based on a scalable multi-disk, multi-node server and an MPEG-4 capable client. We introduced a mechanism to stream variable bit rate media in a simple yet flexible way via the industry standard protocols RTP and RTSP. We demonstrated the feasibility of streaming near NTSC-quality video and audio (compressed via MPEG-4 and MPEG-1 Layer 3 algorithms) to residential locations over current broadband connections (ADSL).

In the near future, we plan to scale up our prototype to more nodes and evaluate its scalability and fault-tolerance with a large number of different clients and media types as well as enable multiple retransmissions.

Acknowledgments

The authors would like to acknowledge the many suggestions and implementation help of Christos Papadopoulos and Rishi Sinha from the University of Southern California.

References

1. J. Al-Marri and S. Ghandeharizadeh. An Evaluation of Alternative Disk Scheduling Techniques in Support of Variable Bit Rate Continuous Media. In *Proceedings of the International Conference on Extending Database Technology* (EDBT), Valencia, Spain, March 23-27, 1998.
2. S. Berson, S. Ghandeharizadeh, R. Muntz, and X. Ju. Staggered Striping in Multimedia Information Systems. In *Proceedings of the ACM SIGMOD International Conference on Management of Data*, 1994.
3. E. Chang and H. Garcia Molina. Efficient Memory Use in a Media Server. In *Proceedings of the International Conference on Very Large Databases*, 1997.
4. J. Hibbard. What the $%@# is DivX;-)? *Red Herring Magazine*, January 2001.
5. Ch. Papadopoulos and G. M. Parulkar. Retransmission-based Error Control for Continuous Media Applications. In *Proceedings of the 6th International Workshop on Network and Operating Systems Support for Digital Audio and Video* (NOSSDAV 1996), Zushi, Japan, April 23-26 1996.
6. V.G. Polimenis. The Design of a File System that Supports Multimedia. Technical Report TR-91-020, ICSI, 1991.
7. J. R. Santos and R. R. Muntz. Performance Analysis of the RIO Multimedia Storage System with Heterogeneous Disk Configurations. In *ACM Multimedia Conference*, Bristol, UK, 1998.
8. H. Schulzrinne. RTP: A Transport Protocol for Real Time Applications, 1996. URL: http://ww.itef.org/rfc/rfc1889.txt.
9. C. Shahabi, G. Barish, M.R. Kolahdouzan, D. S.-Y. Yao, R. Zimmermann, K. Fu, L. Zhang. Alternative Techniques for the Efficient Acquisition of Haptic Data. In *Proceedings of the*

SIGMETRICS 2001 / Performance 2001 conference, Cambridge, Massachusetts, June 16-20, 2001.

10. F.A. Tobagi, J. Pang, R. Baird, and M. Gang. Streaming RAID-A Disk Array Management System for Video Files. In *Proceedings of the First ACM Conference on Multimedia*, pages 393–400, Anaheim, CA, August 1993.

11. R. Zimmermann and S. Ghandeharizadeh. Continuous Display Using Heterogeneous Disk-Subsystems. In *Proceedings of the Fifth ACM Multimedia Conference*, pages 227–236, Seattle, Washington, November 9-13, 1997.

QuDAS: A QoS-Based Brokering Architecture
for Data Services

Nektarios Georgalas

Btexact Technologies Research, Adastral Park B54/Rm125, Martlesham Heath, Ipswich,
IP5 3RE, UK
Nektarios.Georgalas@bt.com
http://www.labs.bt.com/people/georgan

Abstract. Telecommunications companies currently present a large demand for
flexible and efficient data management system architectures. Data Management
Systems are providers of data access services, or simply *data services*, which
perform functions that manage, manipulate or deliver data. In order for such
systems to provide high quality data services that adequately meet the special
requirements of telecommunications companies, they should comprise the following dimensions: (i) the data service users (humans or client applications)
should be granted sufficient control to select data services according to their
specific Quality of Service (QoS) requirements (ii) the provided data services
must present explicit descriptions of their non-functional –quality- properties
along with their functional characteristics, and (iii) dynamic mechanisms should
be adopted for adaptive data service resource management in order to achieve
the best of performance under any circumstances. Recent developments in the
areas of data management and database middleware fall short in proposing approaches for data service delivery that emphasise these dimensions and especially the QoS aspect of a data service. In this paper we present QuDAS, a QoS-
based data access provisioning system that is strictly designed on the lines of
the above criteria. The main contribution of QuDAS is that data services are capable of publishing their functional *and* quality properties as a special set of
meta-data, namely, *interfaces*, *protocols* and *tariffs*. Based on this meta-data users can select any set of data services that is tailored to their functional *and* QoS
needs. Additionally, QuDAS achieves to dynamically manage service resource
usage and automatically balance its workload by means of tariffs, which are
continuously adjusted driven by the varying user demand for data services.

1 Introduction

Telecoms[1] service management and provisioning have currently become very data-
intensive processes because of new developments in information and telecommunication technologies. For example, itemised billing requires a detailed account of customer call data that ought to be collected from the switches; in intelligent networks

---

[1] Abbreviation for *telecommunications*

W. Jonker (Ed.): Databases in Telecommunications II, LNCS 2209, pp. 126–139, 2001.
© Springer-Verlag Berlin Heidelberg 2001

(IN) real-time data access is instrumental for the reliable performance of special IN services like call divert; personalising internet services requires real-time exploitation of data collected in the ISP caches where the customer's behaviour is recorded. Therefore, Data Management Systems (DMS) are greatly important for the efficient delivery and management of telecommunication services. Further challenging data management requirements are introduced when considering the massive volumes of data generated by telecoms-specific processes, the distribution of sites this data is collected and the need to integrate local DMSs for the delivery of complex data management services. Consequently, DMSs in telcos[2] should be effective, flexible, open and capable of providing for efficient use of *data services*.

Data services are services that manage and manipulate data. These can be delivered by a platform of one or a combination of several distributed DMSs, which have the role of the data service provider. A key aspect for the effective delivery of services by the provider is efficient resource management to achieve high levels of performance. This should involve automatic mechanisms that dynamically regulate resource usage towards keeping a right balance. In parallel, another key aspect for the effective use of data services is to provide users (either humans or other client applications) with the power to select the set of services that meet their requirements. Specifically, users should have a high degree of control over determining the levels of service *quality* to be received by the data service provider. To achieve both aspects, some sort of sophisticated database middleware is in need that manages the access to data services with no compromises to requirements set by either the provider or the user.

Most of work in the area of database middleware deals with the integration of different data sources that are distributed over a computer network. There are two main approaches to this problem. The first is the Federated DMS approach which describes a DMS with a uniform view of data achieved through the adoption of one global data model over the local data models of the component data sources. The second approach is based on mediator systems, which utilise the functionality of wrappers to access and translate the information from the local sources to the global data model. Examples of systems that implement these approaches are TSIMMIS [22], Garlic [21] and DISCO [20]. However, all these approaches provide data- and not service-oriented solutions. That is, they mainly focus on how to structure and query data to achieve efficient service performance but they miss the point of examining the QoS aspect for the provision of data services.

The concept of QoS was initially established in the area of multimedia and network communication services. Several systems and architectures have been proposed to guaranty delivery of such services at a QoS level that users and providers have agreed upon in advance. Some examples are [3] and [4]. QoS was recently introduced in the specification of higher-level services, which are delivered by software components. Here, QoS specifications explicitly describe non-functional characteristics of software systems such as reliability, performance, timing and security and have the form of a contract that binds users and the system components, which deliver the service. It is necessary that the middleware situated between users and service providers is extended with special mechanisms that maintain the QoS contract valid during service delivery time. An example language for component QoS specification is QML [6] and

---

[2] Abbreviation for *telecommunications company*

an example system that provides QoS support for middleware is QuO [18]. Nevertheless, the introduction of QoS in the area of data management is only just beginning to receive attention by the research community ([2], [13], [7] and [12]).

We consider that QoS in the particular area of data services is influenced by two factors: the *quality of the resource* used by the service and the *quality of the implementation* by which the system delivers the service. The main resource of data services is data. Therefore, the quality of the data becomes important. Some indicative work on data quality can be found in [4], [11], [9], [15], [16] and [17]. The quality of the implementation is a factor mostly relevant to the performance of the data service system. Generally, data systems do not explicitly describe their QoS parameters. Instead, they implicitly assume these parameters when they are designed so that their implementations can guaranty the assumed quality will be achieved.

In this paper we present QuDAS[3], a system that provides access to data based on the specification of data service QoS the users require. The main contributions of QuDAS are three. First, data services are capable of publishing their functional and quality properties as a special set of meta-data, namely, interfaces, protocols and tariffs. Second, based on this meta-data users can select any set of data services that is tailored to their functional *and* QoS needs. Third, QuDAS achieves to dynamically manage service resource usage and automatically balance its workload by means of tariffs, which are continuously adjusted according to the user demand for data services. The latter feature is encountered in adaptive resource management and service allocation approaches, such as [8] and [14], that apply micro-economic and market models. A DMS that applies a similar approach is MARIPOSA [19].

A first prototype implementation of QuDAS was developed for experimentation purposes of the EURESCOM P817 project [5].[4] Further development of key QuDAS ideas, not particular for data but for general application services, is currently undertaken in the ANDROID project [1].[5]

The rest of the paper is structured as follows. First, an overview of the QuDAS architecture is presented. Next, each QuDAS component is specifically emphasised. Subsequently, we describe how QuDAS works and how it manages the use data services. Finally, we close with discussion and further work.

---

[3] QuDAS stands for Quality of service based system for DAta Services

[4] "Database Technologies for Large Scale Databases in telecommunications", EURESCOM project P817

[5] "Active Network Distributed Over Infrastructure Development", ANDROID, Framework V.

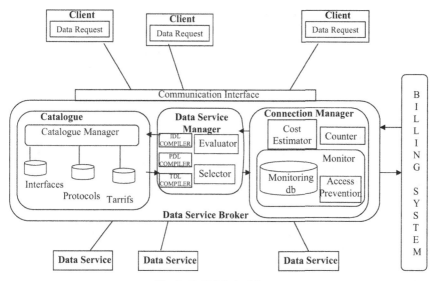

Fig.1. QuDAS Architecture

2   QuDAS Overview

The QuDAS system architecture is illustrated in **Fig.1** and consists of the following modules:

- *Data Services* which provide data access functionality,
- the *Broker* that consists of:
  - a *Communications Interface* which receives data access requests from client applications,
  - a *Catalogue* which is configured for storing an identifier for every data service available. Additionally the catalogue includes two types of information for each data service. First is information about the data access functions implemented by the service and second information about the service resources relevant to every data access function the data service implements,
  - a *Data Service Manager* with two parts, namely, the *Evaluator* and the *Selector*. The former is responsible for comparing a received data access request, which includes a data access function requirement and a data service resource requirement, with similar information stored in the Catalogue, in order to identify data services capable of accessing data in accordance with the request. The latter selects a data service identified by the evaluator for data access,
  - a *Connection Manager* which connects clients with data services and monitors the usage of data services.
- *Clients* which send requests for data services that meet specific requirements

3   Data services meta-data

Each data service is characterised by 3 types of meta-data: interfaces, protocols and tariffs.

3.1 Interfaces

An *interface* specification describes the functional data access operations the data service is capable of executing. An interface specification may be defined using a standard language such as the CORBA Interface Definition Language (IDL) or Java. An example interface for retrieving customer call data expressed in JAVA is:

```
interface customer_call_data {
  public CustomerDetails getCustomerDetails (int
cust_id);
}
```

The interface hides the implementation of the getCustomerDetails() method by only describing the format of the request. The implementation that carries out the request can be any algorithm, for example, a sequential search through the data, an index search based on cust_id or a search request to a system operator. Therefore, the performance of the data service may be variable. Considering that service performance has effects visible to the client and that interfaces are not adequate to present to clients such service aspects we introduce a special type of meta-data called *protocol*.

3.2 Protocols

A *protocol* describes non-functional and resource-related aspects of a data service. In this regard, a protocol specification defines values for QoS parameters[6], for example, data service response time, values for data accuracy, data correctness, time since last update, etc. A *protocol* is a statement of the characteristics of an interface instance. That is, it does not define the characteristics of the interface, but rather it defines the characteristics exhibited by an interface when combined with a service implementation. As such an interface may have more than one protocol. The protocols are expressed in a special language called the "*Protocol Definition Language*" (PDL). An example of a protocol for the customer_call_data interface is:

---

[6] Note that data service QoS, as stated in the introduction, involves service performance and quality of data

```
protocol summary_data describes customer_call_data {
  CustomerDetails getCustomerDetails (int cust_id) {
    accuracy == 0.9
    min_execution_time == 100
    max_execution_time == 1000
    timeliness == 24
  }
}
```

The parameters specified in a protocol depend on the nature of the data service to which the protocol relates. Each data service periodically monitors these parameters to get their current values and determine whether the protocol is continuously met.

3.3 Tariffs

A *tariff* specification describes the cost of a protocol. In other words, the tariff shows the cost of executing a data access function in an interface while using a particular protocol. Additionally, it specifies the resource limitations imposed on the data service associated with the tariff. These limitations involve the number of client connections the data service can accept for that tariff. For a data service to support a tariff it must provide an implementation of the interface associated with the tariff, which delivers the characteristics described in the protocol of the tariff. The tariffs are defined in the "*Tariff Definition Language*" (TDL). Here is an example:

```
tariff summary_data_cost on summary_data {
  CustomerDetails getCustomerDetails (int cust_id) {
    resource_usage == 10
    cost_per_execution== 200
  }
  max_resource_usage == 100
}
```

In the example, the getCustomerDetails method costs a client 200 units each time it is called and adds a 10-unit method execution cost on the service's workload characterising the usage of data service resources by the method for delivering its operation. This tariff has been allocated 100 units of total resource usage, as a maximum that should not be exceeded. Therefore, given that only 1 method call per connection is permitted, this tariff can support a 10 connection maximum which results from the ratio of *max_resource_usage:resource_usage*. The Data Service Broker and the Billing System use the cost_per_execution information to charge client applications for using a data service and also to select the lowest cost data service that can meet a client's requirements. The Connection Manager uses the parameters resource_usage and max_resource_usage in order to determine the resource capacity available to the data service to accept more client connections. The cost_per_execution and resource_usage parameters are defined separately as some data services may not impose large workloads on the underlying database systems but their use may be costly for other reasons, for example because a significant amount of pre-processing is required to achieve the accuracy specified in the respective protocol. Tariffs are defined separately from protocols so that data services can support the same protocol at a variety of different costs depending on their respective workload and data access implementation techniques.

4. Broker

The Broker is the core component of QuDAS. It comprises three sub-components that we analytically examine below.

4.1 Catalogue for service meta-data

The Catalogue acts as the main repository where the broker keeps data service metadata, namely, interfaces, protocols and tariffs. Each is stored in specific databases within the Catalogue.

4.1.1 Interfaces storage

This database stores all interfaces implemented by the available data services. Each data service provider, i.e. database server, grants the information about interfaces its services support. An interface record includes the interface name, the identity of the database server that submitted the interface, the identity of the data service or services the interface relates to and details relating to the interface definition including the interface definition source code. The name of each interface is also registered in the Catalogue Manager.

4.1.2 Protocols storage

This database stores the protocols of each data service. More specifically, each database server publishes protocol specifications for the data services it provides. A protocol record includes the protocol name, the identity of the database server that submitted the protocol, the identity of the data service or services the protocol relates to and details relating to the protocol definition including the protocol definition source code. The name of each protocol is also registered with the Catalogue Manager. The protocols storage, additionally, includes an entry for each protocol QoS parameter containing the name of the protocol it belongs, the name of the data access function to which the parameter applies, the name of the parameter and its value.

4.1.3 Tariffs storage

This database stores the tariffs for each data service. Each database server publishes one or more tariff specifications for each data service it provides. A tariff record includes the name of the tariff, the identity of the database server that submitted the tariff, the identity of the data service the tariff relates to and details relating to the tariff definition including the tariff definition source code. The name of each tariff is also registered in the Catalogue Manager. Additionally, the database includes an entry for each parameter defined in a tariff including the name of the tariff, the name of the data access function to which the parameter applies, the name of the parameter and its value.

4.2 Data Service Manager

The Data Service Manager is connected to the Catalogue for identifying data services that are capable of meeting the user requirements. Each client application is provided with a data request means through which data access requests are issued to the Broker. Requests are received at a Communications Interface, which is connected to the Connection Manager. The Connection Manager connects client applications to appropriate data services selected by the Data Service Manager. The Data Service Manager comprises an IDL compiler, a PDL compiler and a TDL compiler for querying the respective databases in the Catalogue. Alternatively, the compilers may be provided to the client applications to allow them direct access to the service meta-data storages. An Evaluator is provided in the Data Service Manager for comparing received data access requests from client applications with the interface, protocol and tariff specifications in the Catalogue. The Evaluator identifies the data services that are capable of meeting the requirements of the request. Additionally, there is the Selector component in the Data Service Manager, which is programmed to select the most appropriate data service identified by the Evaluator. This selection is based on a pre-determined selection strategy that points to the data service with the lowest cost.

The Data Service Manager implements a number of programming interfaces for selecting an appropriate data service including: i) a broker_access interface for locating appropriate data services ii) a protocol_definition interface for querying protocols and iii) a tariff_definition interface for querying tariffs. We give examples of these interfaces using Java:

```
interface broker_access {
  String []findDSInterface( String interfaceName );
  String []findDSProtocol( String protocolName );
  String []findDSTariff( String tariffName );
  String getDSInterface( String serviceAddress );
  String getDSProtocol( String serviceAddress );
  String getDSTariff( String serviceAddress );
}

interface protocol_definition {
  String[] getProtocolNames();
  String[] getProtocolMethods( String protocolName );
  String[] getProtocolParameters( String protocolName,
String methodName );
  Object getProtocolParameterValue( String protocolName,
String methodName, String paramName )
}
```

```
interface tariff_definition {
  String[] getTariffNames();
  String[] getTariffMethods( String tariffName );
  String[] getTariffParameters( String tariffName, String
  methodName );
  Object getTariffParameterValue( String tariffName,
  String methodName, String paramName );
}
```

A client locates an appropriate data service using find data service methods of the borker_access interface, for example:

```
String [] service = broker.findDSTariff( tariffName );
```

Each findDSX[7] method returns one or more available data service addresses. Such an address looks like a standard Java RMI connection address. If more than one data service is available the client application can query the Catalogue's databases by passing the service address to the getDSX methods of the broker_access interface to obtain the name of the respective interface, protocol or tariff. Additionally, the client application may query the Catalogue about the properties of a named protocol or tariff using the protocol_definition and tariff_definition programming interfaces.

4.3 Connection Manager

The Connection Manager comprises a Monitor, which is connected to a Monitoring Database. The Monitor controls the current usage of all the data services listed in the Catalogue. The Monitoring Database stores one record with name and address of each data service registered with the Broker. The database is also provided with a counter, which maintains a count of the current usage of each data service by clients. The Connection Manager uses this information to determine whether a preferred data service is capable of supporting further connections. If the pre-determined maximum level, as indicated by the *max_resource_usage* tariff parameter, has been reached subsequent connections will be prevented by an access prevention means. If the connection is permitted then it gets established and a relevant data service access event is recorded in the Monitoring Database. A similar event is stored when the client request for connection has been prevented or an established connection ends.

The Monitoring Database is further associated with a Cost Estimator mechanism. This updates the Catalogue's tariffs database adjusting the *cost_per_execution* parameter in a data service tariff to reflect client demand for the service. For example, bigger demand may motivate an increase to the tariff cost of a service as a means to control client over-subscription. There are two algorithms the Cost Estimator may follow. The first checks for each incoming request the current resource availability and if there is not enough resources the cost of the requested method increments. The second algorithm checks the number of requests over a time period and if it exceeds a

---

[7] X substitutes for interface, protocol or tariff

particular threshold the cost increments in proportion to the magnitude of the excess. Similar algorithms can be defined for decreasing the cost of a service.

Algorithm1:
```
Tariff t1 = requested tariff
int curRes = total current resource_usage⁸ for t1
int maxRes = t1.max_resource_usage
int minRes = minimum resource_usage found in a t1 method
int diff = maxRes - curRes
if diff < minRes
   for each new connection request
      increment cost_per_execution of the requested method
```

Algorithm2:
```
long T = some constant period of time
long t0 = some starting time instance
long t = current time
int totalRequests, threshold = some default no of re-
quests, N = 1
while (true) {
   if (t == t0+N*T){
      totalRequests = no of requests for (N-1)*T < t < N*T⁹
      int diff = totalRequests - threshold
      if diff > 0 {
         increment cost_per_execution in proportion to the
         magnitude of diff
         N = N+1
      }
   }
}
```

5. Billing data services

The Broker is connected to a Billing System for bill production. The Billing System determines the amount to charge subscriber accounts for the use of data services. This amount relates to the cost of each data service connection and is calculated using tariff data provided by the Catalogue and event data provided by the Monitoring Database.

---

[8] resoure_usage should be the summation of such parameters in all methods supported by tariff t1.

[9] This is the number of events for satisfied or prevented connection requests occurring in the designated time period and recorded in the Monitoring Database

6. Managing access to data services

When a client requires access to data services, it firstly connects to the Data Service Broker, which then asks the client to provide a data service request. The client may prompt the system user for details on a data service through a GUI, for example, with drop down menus of interface, protocol and tariff names. A fully defined data service request should include the name of an interface, a protocol and a tariff. In composing the request, the client application may require access to the Catalogue for the user to run through the published service meta-data. Finally, the data access request is issued to the Broker.

In the Broker, the Data Service Manager determines whether the data access request is fully and correctly defined. Based on the request, the tariff database is searched for data services that exactly meet the client requirements. There, the name and address of one or more appropriate data services are identified. If a set of candidate services are found the client should indicate one specific option. If this does not happen, the Broker automatically selects on behalf of the client the service with the lowest usage cost.

Once a data service has been selected, the Connection Manager accesses the Monitoring Database and queries the Counter to determine the current usage of the service. If the current usage is below a maximum value the Connection Manager calls the data service address and establishes a connection between the client application and the respective database server. Otherwise, the client is contacted in order for it to provide alternative directions e.g. agree on selecting another service or change the request altogether. When connection between client and service is achieved the Connection Manager accesses the Monitoring Database and increments the respective data service usage counter by the value of the *resource_usage* tariff parameter. When the connection ends the usage counter decrements by the same value.

When a data service provider is interested in introducing new data services into QuDAS it has to publish to the Broker all the respective service meta-data. The Broker provides publishX() methods for this process, where X is either Interface, Protocol or Tariff. For example, the method publishTariff (t, DS) publishes a tariff t for a data service DS.

7. Discussion and Further Work

It is shown that interfaces alone cannot present adequate information about the QoS clients should anticipate from data services that implement the interface operations. This leads to the introduction of protocols. Protocols describe non-functional features of a data service. Clients can see in protocols a declaration of QoS property ranges the service is bound to deliver. Hence, the protocol is essentially a form of contract between client and data service provider that guaranties a certain level of QoS. QoS properties in protocols reflect the efficiency and performance of data management techniques, such as distribution, replication, caching, usage of relational or object databases, the data services use. This happens not by directly exposing the technique itself but rather by explicitly describing the technique's impact on the expected QoS.

Tariffs are useful for costing the usage of a data service. They define cost for executing a service operation and cost for occupying resources during service delivery. Cost for execution is directly charged on client accounts. Resource usage cost is a means to determine the current workload on a service plus its capability to accept or reject additional incoming workload. A tariff also specifies the maximum resource usage capability of a service. QuDAS may adjust tariffs after monitoring the client demand for services. Hence tariffs become a dynamic tool to control service over-subscription.

Further work on ideas applied in QuDAS is currently pursued in the ANRDOID project. The latter investigates the management of application services (not strictly data services) in active networks and aims at adopting a similar to QuDAS, but rather more active[10], broker-based architecture that achieves a highly dynamic service provision. Some challenging issues there involve having: (i) a network of active brokers that adapt and behave dynamically in response to environment conditions e.g. detection of increasing client demand may dynamically initiate broker instances situated closer to these clients, (ii) brokers that accommodate client requirements for services with highly customizable solutions that also consider conditions imposed by service providers such as in the case where a client specifies a service but desires to avoid specific providers and simultaneously candidate providers deliver this service but only to certain classes of clients.

In conclusion, it is noted that QuDAS manages to provide benefits for both clients and data service providers. First, clients become entitled to tailor a set of data services that delivers exactly what it is required, neither more nor less, mainly based on specifying the desired QoS levels. This, in real terms, means that customers are billed to get from the selected data services only the value and quality they specify and not the value or quality the service provider coerces them to pay for. Additionally, the use at the service provider end of a cost on service resources that varies according to resource usage and demand introduces a very flexible workload balancing mechanism. This mechanism dynamically adjusts to the conditions posed by the volume of requests for certain data services over time. Both of these conclusions deliver a big advantage especially to telcos whose business market is at present highly competitive.

Acknowledgements

We gratefully acknowledge the key role of Dr Stephen McKearney from the Dept of Computing, Bournemouth University, UK, in developing the ideas of QuDAS. We also thank all students from the same department who involved in the implementation of QuDAS during their short-term placements with BT.

---

[10] Active in terms of its components presenting at execution time not pre-determined but dynamic behaviour which is driven by certain policies.

References

[1] ANDROID project, "Active Network Distributed Over Infrastructure Develolpment", see http://www.cs.ucl.ac.uk/research/android/

[2] Bertino, E., Elmagarmid, A. K. and Hacid, M.: Quality of Service in Multimedia Digital Libraries. SIGMOD Record, Vol 30, No 1 (March 2001)

[3] Campbell, A., Coulson, G. and Hutchison, D.: A Quality of Service Architecture. ACM SIGCOMM Computer Communication Review journal, Vol. 24, No 2, (April 1994)

[4] Dvorak, C., A. and Richters, J., S.: A Framework for Defining the Quality of Communications Services, IEEE Communications Magazine, (October 1988): 17-23

[5] EURESCOM project P817, "Database Technologies for Large Scale Databases in telecommunications", January 1998 – December 2000, see http://www.eurescom.de/public/projects/p800-series/P817/p817.htm

[6] Frolund, S. and Koistinen, J.: QML: A Language for Quality of Service Specification. Hewlett-Packard Labs, Palo Alto, 1998 see http://www.hpl.hp.clm/techreports/98/HPL-98-10.html

[7] Jarke, M. and Vassiliou, Y.: Data Warehouse Quality: A Review of the DWQ Project. Proceedings of the 2nd Conference on Information Quality, MIT, Cambridge US, (1997)

[8] Lalis, S., Nikolaou, C., Papadakis, D. and Marazakis, M.: Market-driven Service Allocation in a QoS-capable Environment. Proceedings of 1st International Conference of Information and Computation Economies, Charleston SC, USA, (October 1998)

[9] Levitin, A. and Redman, T.: Quality Dimensions of a Conceptual View. Massachusetts Institute of Technology (MIT) Sloan School of Management, Cambridge, MA, TDQM-95-04, (February 1995)

[10] Rodriguez-Martinez, M. and Rousopoulos, N.: MOCHA: A Self-Extensible Database Middleware System for Distributed Data Sources. Proceedings of SIGMOD 2000, Dallas, Texas SU, (May 2000)

[11] McKeating, A.: Quality Can Stop Dirty Data. Computerworld, Vol. 26, no. 49, (1992): 33

[12] Mihaila, G., A. and Raschid, L.: Querying Quality of Data Meta-data. Proceedings of the 3rd IEEE Meta-data Conference, Maryland, USA, (April 1999)

[13] Naumann, F., Leser, U. and Freytag, J., C.: Quality-driven Integration of Heterogenous Information Systems. Proceedings of 25th International Conference on Very Large Data Bases, Endiburgh, UK, (September 1999)

[14] Stratford, N. and Mortier, R.: An Economic Approach to Adaptive Resource Management", Proceedings of the 7th Workshop on Hot topics in Operating Systems, IEEE Computer Society, Rio Rico, Arizona US, (March 1999)

[15] Strong, D. and Wang, R.: Beyond Accuracy: What Data Quality Means to Data Consumers. Massachusetts Institute of Technology (MIT) Sloan School of Management, Cambridge, MA, TDQM-94-10, (October 1994)

[16] Wand, Y. and Wang, R.: Anchoring Data Quality Dimensions in Ontological Foundations. Massachusetts Institute of Technology (MIT) Sloan School of Management, Cambridge, MA, TDQM-94-03, (June 1994)

[17] Wang, R. and Madnick, S. "Introduction to the TDQM Research Program", see http://web.mit.edu/tdqm/www/intro.html

[18] Zinky, J., A., Bakken, D., E. and Schantz, R., E.: Architectural Support for Quality of Service for CORBA Objects. Theory and Practice of Object Systems journal, Vol. 3, No 1, (1997)

[19] Stonebraker, M., Aoki, P., M., Devine, R., Litwin, W. and Olson, M.: Mariposa: A New Architecture for Distributed Data. Proceedings of 10th IEEE International Conference on Data Engineering, Houston, Texas USA, (February 1994)

[20] Tomasic, A., Rashid, L. and Valduriez, P.: Scaling Heterogeneous databases and the Design of DISCO. Proceedings of the 16th International Conference of Distributed Computing Systems, IEEE Computer Society, Hong Kong, (May 1996)

[21] Roth, M., T. and Schwarz, P.: Don't Scrap it, Wrap it! A Wrapper Architecture for Legacy Data Sources. Proceedings of 23th International Conference on Very Large Data Bases, Athens, Greece, (September 1997)

[22] Garcia-Molina, H., Papakonstantinou, Y., Quass, D., Rajaraman, A., Sagiv, Y., Ullman, J., Vassalos, V. and Widom, J.: The TSIMMIS approach to mediation: Data models and Languages. Journal of Intelligent Information Systems, (1997)

LDAP, Databases and Distributed Objects: Towards a Better Integration

Thierry Delot*, Pascal Déchamboux, Béatrice Finance*, Yann Lepetit, Gilles LeBrun

France Telecom R&D Lannion Grenoble
*PRiSM Laboratory University of Versailles St Quentin
emails : {Thierry.Delot, Beatrice.Finance}@prism.uvsq.fr
{Pascal.Dechamboux, Yann.Lepetit, Gilles.Lebrun}
@rd.francetelecom.com

Abstract. For the needs of web platforms (portals, e-commerce, telephony on Internet, etc) and network platforms (switches, routers, SCPs, etc), organizations manage more and more data using LDAP servers. LDAP (Lightweight Directory Access Protocol) [20] is the standard, proposed by the Internet Engineering Task Force (IETF) for modelling and querying network directory information, as well as accessing network directory services. It also provides a set of services to manage authentication and security. For web and network applications, both LDAP servers and traditional databases are now used and will continue to be used, leading to big interoperability issues. As stated by IETF [13] we are faced to the integration of LDAP and Database technologies to provide a "highly distributed and scalable network database service". Moreover, in the management field, DMTF is promoting LDAP as a key technology to manage heterogeneous network nodes, leading to another integration challenge; how to provide a "LDAP view" on objects. In this paper, we analyze these new requirements and overview some of the studies currently done by France Telecom and PRiSM to better integrate LDAP and Databases technologies, as well as to integrate LDAP and OMG distributed objects.

1 Introduction

During the nineties, the telecom community was strongly influenced by the object technologies proposed by ODP, OMG and now EJB. These technologies were supposed to answer to all the computing issues in the telecom world and to mask the database issues. Several attempts [4,10,21,24] were made to smoothly integrate the data models and data queries within the OMG/ODP model. *Is this integration effort enough to match all the needs?*
Indeed, distributed object technologies are important for the pure telecom world, but since mid nineties W3C/IETF is also of big importance. The telecom world has been

W. Jonker (Ed.): Databases in Telecommunications II, LNCS 2209, pp. 140–154, 2001.

strongly impacted by the IP wave. More and more communication services are transferred into HTTP servers re-grouped into portals. These IP portals have now a big chance to become the service layer of the so-called Information Network. But, from a technology viewpoint, W3C/IETF is much more pragmatic than OMG. It relies as much as possible on simple technologies based on lightweight protocols: HTTP, LDAP, SIP, etc. In that way distributed applications can be weakly coupled and can be implemented with heterogeneous software technologies (i.e. heterogeneous OS, DBMS, ORB, J2Eplatforms, etc). In this context, LDAP is very important due to its simplicity and its wide diffusion within the Internet world.

Like this, LDAP could play a very important role in a seamless Information Network integrating New Generation Network (NGN) based on IP and IP portals hosting most of the communication services. LDAP is used for users and services directories within WEB platforms (portals, e-commerce, etc). It is used within IP transport networks (routers) as resources directory. The Distributed Management Task Force (DMTF) [8] plans to use it as resources and services directory for private networks. As a consequence, within the Network Data Management field, LDAP should become the most "lowest common denominator" to different heterogeneous technologies and services that could cooperate by sharing information through LDAP. Naturally, LDAP will not solve alone all the interoperability issues in Information Network, but it can become a strong federating element for the data aspect, as IP is for the communication aspect.

However if the LDAP protocol is well-suited for Information Networks, from a data management viewpoint, LDAP servers have some lacks and DBMS are necessary for applications requiring high-level database services. We are then faced to the integration of LDAP and Database technology to provide a "highly distributed and scalable network database service" as advocated by IETF [13], such a service is being considered as a "central architectural issue". Another important issue, but more academic as it may have big impact on the standard, is to enhance the querying and browsing capabilities of LDAP [18]. All these objectives pose big challenges that are presented in this paper. Starting from current and future requirements, we analyze the research issues we address and outline some general solutions that we are beginning to explore.

2 Problematic and Requirements

In this section we overview the needs of web platforms (portals, e-commerce, telephony on Internet, etc) and network platforms (switches, routers, SCPs, etc) and the technical limitations we are facing today to support them. What is clear is that these systems will manage more and more data using LDAP and despite the fact that up to now LDAP servers fulfil some needs, they will face new challenges such as the followings :

- **Scalability, QoS and Performance**

Future directories will have to manage very large bases that will frequently reach tens of millions of entries. For example, the directory of France Telecom contains up to 20 millions of entries and will still increase in the future. Today, lots of internet users are doing web connections during the middle of the night and for very long sessions (i.e.

many hours) and they don't accept to be interrupted for fault or night network maintenance reasons. As a consequence, we must now provide high availability transport network services and use non stop network databases. Some services such as mobile phone use very stringent real time databases (HLR), remotely accessed with the MAP protocol on top of the SS7 ITU protocol. With the evolution toward the NGN (New Generation Network), network architects would like to substitute LDAP on IP to MAP, but the real time constraints are very stringent (less than a millisecond for a remote response time).

Today's commercial LDAP servers have difficulties to match such stringent requirements. They have usually their own lightweight physical data layer that presents several drawbacks and which can not fulfil certain constraints faced by telecom companies. They offer a limited physical store, usually one disk, compared to multiple disks managed by relational and object base management systems. They have relatively bad performances when dealing with large bases. There is no query optimiser, vertical partitioning techniques on attribute values are not always the best. They offer very weak durability properties (i.e. log must be explicitly managed by users to restore a broken base), asynchronous replication mechanism which does not provide a good availability and consistency, a static administration which often requires to stop the LDAP server. Many of these drawbacks have been already addressed by the database community.

As stated by IETF [13], it is interesting to investigate whether directory services can be seamlessly integrated (from the point of view of the applications using them) with others forms of storage and retrieval (such as relational databases) in order to provide an integrated directory service with these capabilities.

- **Manipulate common data with different views**

Telecom companies (Telcos) are also facing a big competition in WEB services field. To reduce the time to market and differentiate their offer, they combine on the same portal, off the shelves services and Telco designed services [27]. These different services, located on the same platform, must very frequently share common data like authentication data, some part of user profile data, accounting data, etc. Off the shelves services are "black boxes" that impose their own data model to manage their persistent data (Net Meeting or Mail services require an LDAP server for example). Some other services require large databases with transactions and high availability that can only be provided by relational DBMS.

It is then more and more frequent that different applications on the same platform have to manipulate common data but with a different view: a LDAP view and a DBMS view. As a consequence, common data are manipulated with different data servers: relational server and LDAP server for the time being. Usage of two servers, a DBMS and a LDAP server presents a lot of drawbacks: applications must manage the consistency between the two bases, both LDAP and DBMS licences are very expansive, quality of service provided by LDAP is sometimes weak, application developers don't like to administrate two data managers on the same platform.

Here also, it is interesting to offer an integrated server, but providing access to both models.

- **Loosely coupled servers**

Unfortunately, replication of data can not always be avoided. If it is interesting, within a portal or a platform which can be a strongly coupled distributed system (based on Corba or rmi), to centralise the data in an integrated server as explained previously; at a wide area network level it is usually too restrictive. Hence, certain telecom companies provide a wide portal offer (general portal, free portal, specialised portals: education, games, etc). Each portal provides a bunch of services, using its own database managed by a DBMS or a LDAP server. But service subscription is now directly managed in line by the portals in an autonomous way. As a consequence, a user can subscribe asynchronously in several portals, providing each time almost the same information (user profile) that is duplicated in this various portals. Portals are usually managed by administrative entities that want to stay as much as possible autonomous, they don't want to share a common database server. As a consequence, the consistency of replicated profiles is not managed. This induces inconsistency and impacts the final quality of service (QoS) provided to the final user. It leads to problems of convergence of copies which are not easy to deal with and that are, in the best case, left to the charge of the application developer. Moreover corporate marketing people have a lot of difficulties to get a "global view" of a customer, with all its subscribed services and consumer practices.

In all its generality, we are faced to a weak federation problem between DBMS and LDAP servers spread over a wide area network, with two main issues to be solved: consistency of replica and translation of data schema.

• Dealing with distributed objects

ORBs and J2E platforms are also more and more used within switching nodes or portals. The attempts to provide a CMIS like protocol to manage such platforms are not successful, despite their technical credibility [10,11] and LDAP seems to be a good candidate to answer these management needs. The Distributed Management Task Force (DMTF) which groups the main software editors (Microsoft, IBM, SUN, etc) and big networks players (Cisco, Ericsson, etc), is strongly promoting the usage of LDAP for management purposes, via the Directory Enable Networks approach [7]. The idea is to favour the cooperation of networks elements: each network element "publishes" via LDAP, a view on its resource state and service data that can be used by other network elements. These LDAP information can be used for pure management purposes, but also to set up communication services (network policies, communication resource allocation).

Some telecom manufacturers also think that the CMIP protocol (which is usually used in the telecom administration field) is too complex (due to ISO stack and GDMO model) and that SNMP, which is now widely used, has a too low level. They think that LDAP is a good compromise between the two. Some other people believe that LDAP directories can be used not only to store standard objects (i.e. users, organisations, etc), but also to bind network services to their clients.

In parallel to this strong tendency to use LDAP as a management protocol between loosely coupled platforms, there is also an increasing interest in building network applications, located within a distributed platform, upon Corba (Common Object Request Broker Architecture), the distributed object infrastructure promoted by OMG (Object Management Group [23]). This interest is due to the fact that it is much easier to build strongly coupled distributed systems using interoperability technologies at the API level (e.g. Corba).

All that leads to propose a directory service over a Corba platform, for both remote management needs and universal binding not provided by the current Naming Service [25] and Trading Service [26]. These requirements lead to manage much more dynamic information within the LDAP server, the existing LDAP servers do not provide such functionality. The main issue is that network and service objects strongly differ from other directory or data objects; standard directory objects are simple and exist within static boundaries whereas network elements and services are complex objects existing in an evolving environment.

3 Research issues

To answer to the problems presented above. Four main research directions have been identified. The first one consists in taking the best of what directories and database have to offer. This would result in a highly scalable and available database that can use containment to effectively organize and scope its information. It would be able to have exceptional read response time, and also offer transactional and relational integrity. It would support simple and complex queries. The second one consists in offering several views providing different data models of the same data. The third one addresses the problem of replication in a loosely coupled environment. Finally, the fourth one investigates the management of distributed objects.

- **Building a scalable directory service**

As stated above LDAP will be the common denominator. Depending on the target application, several storage can be considered for building a directory service from classical DBMS, to high availability DBMS with sophisticated replication mechanism, to main memory DBMS, or files, etc. Until now, most of the LDAP commercial products have their own lightweight physical data layer that offers good read response time and can be easily embarked on lightweight systems (small network equipments, terminals, etc). However, they often present several drawbacks when dealing with large databases that require a lot of quality of services such as durability, replication, multi-disks, etc. Our approach consists in using DBMS as a storage manager to offer the required quality of services, while keeping the LDAP protocol unchanged. The approach is very similar to what is done in the database field; JDBC drivers are proposed to clients to enable them to write code independently from the database.

We want to investigate relational and object DBMS as target storage managers. Object DBMS are taken into account since they are introduced more and more in the telecom field, following object languages and distributed object technologies usage, such as Corba, RMI, EJB. They provide a seamless object language binding, a good level of performances for tree structures. They are more suited to 3 tiers architecture proposed by distributed object technology [21] and can be more easily embedded than relational DBMS. For the time being, Oracle proposes a Directory Service, but none is available on object oriented DBMS. To our knowledge some experiments have been done on top of VERSANT by Lucent, however no result has been communicated.

Competing with Oracle is not our objective, rather we would like to propose an *open and generic architecture* that can be adapted to any kind of DBMS. Indeed, building a directory service is a complex task and we want to factorize, as much as possible, the development and reuse of code. We also want to control the code, in order to adapt it to needs of certain applications that require, for example, a double access on common data. This is currently not possible with Oracle Directory Server that proposes a 'black-box'; LDAP data can not be accessed via SQL.

Our target architecture, depicted in figure 1, relies on the standard architecture of a LDAP server as adopted by OpenLDAP[1]. Our LDAP server is composed of two main parts : (1) a front-end, which is responsible for clients connections to the server by means of the protocol LDAP, and (2) a back-end which manages data storage as well as queries on these data. The back-end is similar to a specific DBMS which implements the LDAP model. The back-end is constituted of a schema manager, a query evaluator (parser, analyzer, decomposer and query optimizer), a storage manager and a cache. The front-end is in charge of multi-threading and encoding/decoding the messages exchanged between the client and server which are represented with the BER (*Basic Encoding Rules*) format.

As the interface between the front-end and the back-end is clearly specified in OpenLDAP, we suggest to use an existing LDAP front-end to manage LDAP connections. This solution offers certain advantages. It allows to use the numerous LDAP clients, already existing, which proposes a wide variety of language API[2] to access an LDAP server via a standard LDAP protocol. This front-end will not interact with a classic LDAP back-end as it is usually the case, but will be connected with a new back-end. The interface between the front-end and the back-end must naturally follow the commonly agreed interface which mainly consists of the *add, delete, modify, rename, search* and *sendresults* primitives.

Fig. 1. *The architecture.*

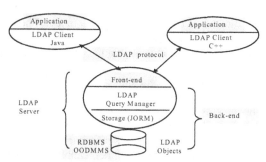

In order to provide genericity, logical models are specified at the query manager level and mapped onto a particular persistence service: JORM (Java Object Repository Mapping) [3], developed as an open kernel for a wide range of Java persistence needs (EJB, rmi, corba platforms). JORM masks the heterogeneity of the persistent models being used for LDAP data, by offering an extensible persistent JAVA mapped onto relational or object DBMSs. While designing this new directory server, we have taken into account many improvements. For example, the new Directory Service is dynamically administrated through a graphical interface avoiding the interruption of the directory server to perform schema modification, as it is the case for LDAP

---

[1] The open source implementation of OpenLDAP is available at http://www.OpenLDAP.org/.

[2] Netscape Clients to connect to Java, JavaBeans, JavaScript, EJB, applications Microsoft ADSI Clients to connect to ActiveX, DCOM, JScript, Visual basic, VBScript, applications.

products. Moreover, since the administration layer of Information Networks will more or more rely on XML, it is important to provide a gateway between LDAP and XML like DSML [9]. The prototype currently being developed supports dynamic schema administration which can be exported in XML following the DSML definitions, query results can be obtained either in LDIF or XML, as it is described in section 4.

- **Manipulate common data with different views**

As stated above, it is more and more frequent that different applications on the same platform have to manipulate common data, but with a different view: a LDAP view and a DBMS view. Two approaches can be followed. The first one exploits the fact that LDAP data are already stored in a DBMS which proposes the view concept and allows several views of the same data. By knowing the logical model of our Directory Service, we define a set of DBMS views providing SQL accesses to LDAP data. These DBMS views can be materialised or not, to offer better performance or to avoid too many interactions between the different accesses via SQL or LDAP. View maintenance, in case of updates, can be offered by the underlying database system that provides very clean update mechanisms based upon triggers. However this approach is satisfying for LDAP servers built from scratch, but does not answer to the needs of applications that require a LDAP access on existing data called "legacy database".

The second approach consists in defining a wrapper as it is usually proposed to federate heterogeneous legacy data sources. This wrapper offers to users, a LDAP vision on legacy relational data. Building a mapping between the relational model and the LDAP model is not easy. It requires some kind of retro-conception. Indeed, these two models are very different. The relational model uses a flat data model, based on tables, while LDAP directories rely on a hierarchical model. The identification of data in the relational model is based on the concept of key; while LDAP uses names which are generated following the containment relationships that the objects have in the hierarchy.

In order to avoid the complexity of generating mapping rules by hand, we propose in [32] a LDAP wrappers generator inspired by existing approaches proposed for extracting XML documents from legacy databases [12, 29]. The solution follows a two-stage approach, i.e. (1) a phase of definition of the wrapper and (2) a phase of generation of the wrapper. This separation is necessary to manage the complexity of the process. To start with, the designer defines simple mapping rules between its relational and a chosen LDAP schema. Then, the system checks the validity of the proposed rules and generates a set of views. Since most of the views are very complex to define and very costly to materialize, the system analyses implementation techniques and selects the one which guarantees the best compromise between data coherence and performances.

- **Consistency in loosely coupled environments**

If we look at consistency for distributed portals, current replication mechanisms provided by DBMS do not fit: replica instances are systematically created following the location information provided by the replication schema that is statically created by database administrators. Here, we need a new kind of replication system enabling to create dynamically a replica, when an instance of the data already exists in the distributed system. Replication meta information must be created starting from data manipulation queries and not from administration queries.

Replication must be managed in a weakly coupled way, with as much as possible consistency. Symmetrical and asynchronous duplication algorithms, as proposed by LDAP, are good starting candidates; but enhancements are needed to match our new requirements. Symmetrical algorithm leads to a decentralised control. It avoids to secure a master server and provides a weaker coupling between servers. It matches our needs. Synchronous update mode, based on a best effort approach, with a degraded asynchronous operating mode in case of server failure, seems the best solution to provide a good level of consistency at low cost.

The main challenge we are faced to, is to design a very simple algorithm that does not make hypotheses on the transactional properties of the underlying storage manager: full transactional DBMS, DBMS or LDAP storage with concurrency management only. As a consequence atomic replication without transaction is only aimed at in a first step, a lightweight transactional model based on one phase commit [28] being considered in a second step. In order to stay as much as possible independent of the underlying servers, the replication manager takes over all the distributed algorithm: global and efficient locking [22], local and global commit, and the replication information (replication schema and localization) are stored in the application data. XA interfaces or local transaction interfaces, when provided by DBMS, are not used. As a consequence, the replication manager can be connected on top of application interfaces of DBMS or LDAP servers.

Another big issue is to determine if a data already existing must be replicated. One solution is to use a unique identifier to determine if the data already exists in the distributed system. This solution presents certain drawbacks: identifier allocation must be secured by complex and fragile distributed allocation algorithms as proposed in [14], knowledge of a system information by final users, management of the allocation history, etc. Another solution, that seems better, is to use application attributes that can constitute a unique key, to determine if a data already exists. In that case, we use information that have a semantic meaning for the user, but a data administrator must define the key for each type of data and the final user must provide values for these attributes.

From an architectural viewpoint, the duplication manager works at a LDAP protocol level, to keep a compatibility with existing LDAP servers. Regarding the schema federation, the schema translation problem can be very complex to solve (as explained in the previous paragraph). Update of translated data can be difficult. More studies are required to provide a good answer in this field. At first, we propose to solve inelegantly the issue by asking applications to modify and define a new common schema for the subset of data that must be replicated.

- **LDAP and distributed objects**

To manage distributed objects, applications often need both naming and trading functionality which provide respectively location transparency and the ability to identify objects relatively to their attributes values. For the time being, these two services are independent and offer functionality which are difficult to couple together. For example, it is not possible to query objects in the same manner as it is done in LDAP. It is of course possible to build a query service on top of these two services as we have already described in a previous paper [11], however this approach is dependent of the inherent limitations implied by the underlying services. Moreover, the deployment of a query evaluator on top of these two services is feasible, but to the prejudice of performances due to optimisation problems. That is why we propose a Directory Service [5] which federates in a clean, open and flexible manner distributed objects. A prototype has been realised in the context of Corba platforms. This Directory Service proceeds in a way very similar to MetaDirectories but it integrates Corba objects instead of DBMS data or protocols as in MetaComm [1].

Several solutions are available to manage Corba objects in a directory. The first one replicates attributes values of Corba objects in the directory, and forwards any object modification to the directory. This solution is very similar to the one used by traders, when clients export information to the trader. However, this solution is not suited to the management of objects with a dynamic behaviour. The second solution avoids replication and instead gives the preference to the proxy. This approach is the one used in the naming service, objects are referenced in the directory information tree relatively to their distinguished names. By using the proxy, users can always access the latest values of the attributes of the referenced Corba object.

From a performance point of view, the solution proposed by the trader provides better response time for querying objects values, since objects values are stored locally. No network access is needed when querying the directory. On the contrary, each time a user or a query service wants to access to an attribute value of an object managed by a naming service, a remote access is performed. This approach is very penalising for objects which do not have a dynamic behaviour.

As explained before, replication introduces inconsistency and therefore can result in bad quality of service for applications requiring high dynamicity. For example, the location property of an object representing a mobile phone may change very often. Moreover, keeping up to date information in the directory for Corba objects having a very dynamic behaviour and with frequently updated attributes, requires numerous updates. This does not always interest users, and have a real impact on performance too. Indeed the directory is always disturbed and for a long period, since updates in a directory are very slow compared to read access. When busy the directory service can not satisfy others demands which have real time constraints.

In [6], we have extended our Corba Directory Service to mainly answer the needs of DMTF and mobile phone applications, which require more flexibility and dynamicity. The properties of the objects managed by such applications are frequently updated, so we avoid solutions only based on attributes replication, which cause both performance and accuracy problems, by providing a transparent and direct access to the properties of named Corba objects. We suggest marrying and widening LDAP technologies and Corba naming service to realize a directory service really adapted to the management of Corba objects with dynamic behavior. The objective is to obtain directories with advanced features able to : (i) treat rich data containing values but also programs, as Corba objects, (ii) preserve LDAP inherent flexibility and (iii) manage in an efficient and transparent way the distribution of the objects referenced in underlying directories.

4 Browsing and Query Facilities

When using a directory, one must also consider any human-oriented usage of a directory. For example, it may not be possible for a user to directly quote a name for the object about which information is sought. However, the user may know it when he sees it. The browsing capability is necessary and will allow a user to wander about the directory, looking for the appropriate objects. Moreover, directory Services usually provide querying facilities to retrieve collections of objects managed in a directory which verify some defined properties. However for the time being, existing management or directory services (CMIS [16], X500 [15], LDAP [20]) follow a protocol approach which is not very convenient for non computer scientists.

For the time being, only LDAP URL offers a subset of search operations that can be used in Web browsers[3] as illustrated in the following. However, no query facilities are offered to add or delete entries in a directory. Moreover, the language is very limited. For example, the query given in figure 2 retrieves all the entries relative to the United-States, with a "CN" attribute value beginning with "Bill".

Fig. 2. *LDAP URL.*

Syntaxe :   ldap://(hostport)/*query_expression*   *query_expression* = dn ? level ? filter ldap://nldap.com/c=us? sub ? (cn = Bill*)

We believe that query languages are needed for these protocols [4] (as already done in the database field with the SQL query language and the RDA protocol [17]). Hence, we think that capitalising on the database technology to propose a query language, while keeping the semantic prescribed by LDAP is an interesting issue. This gives better independence to the program and can allow different kinds of evaluation and optimisation techniques.

Researches have been recently done, within the database community [2,11,18,19], to extend the expressive power of directory languages. In [18], the authors from AT&T propose a formal data model for directories and a sequence of efficiently computable query languages with increasing power. In particular, they define hierarchical selection operators to exploit containment relationships between entries. They propose other functionalities such as aggregate selection operators and a closure property that permits queries to be composed.

In [6], we have proposed a language, called DQL (Directory Query Language). DQL does not only provide query facilities but also statements to create or destroy entries. Since the LDAP query language does not provide the means to answer directly to a query like "Retrieve, for each company, the employees earning more than 5000$". We extend DQL to manage regular path expressions, to query the Directory Information Tree (DIT) accordingly to the containment links that exist between objects. Notice that regular path expressions have been proposed by query languages defined for semi-structured data such as XML (i.e. XPATH [30], XQuery [31]). In LDAP, several queries are needed in order to, first retrieve the distinguished name of each company, then for each one of them retrieve their employees. As containment links between objects are lost in the query result, the user must exploit the Distinguished Name (DN) to retrieve the containment relations that exist between the entries of the directory; nevertheless, this operation, even if it is feasible, is very fastidious and costly.

---

[3] Examples are available on the web (ldap://nldap.com/?cn=admin,o=NOVELL?sub?(sn=g*))

For the time being, LDAP query results are considered as an unstructured bag of entries defined in LDIF (LDAP Data Interchange Format) which is very limited; hierarchic links between entries are lost in the result. We think that a hierarchic presentation of LDAP query results based on XML is very useful for complex LDAP data manipulations. This requires some modifications of the DSML proposal that exports only LDIF files in XML.

The approach we are taken within DQL is to better integrate XML and LDAP query languages. Even if LDAP shares the flexibility of semi-structured data model as mentioned in [19], simulate a Directory Information Tree with a semi-structured data model is possible, but not well-suited because (1) the separation between data and structure will be lost in the simulation, (2) hierarchical naming principles are not taken into account, and (3) scooping is not easy to implement. Apart from that, semi-structured query languages rarely address the problem of the distribution of semi-structured data and the problems they carry with (i.e. difficulty of query decomposition, need of partial answers, unavailability of data sources, tasks scheduling,...). These aspects are studied in the context of DQL.

In the following, we give some screen copies of the browsing facility coupled with the DQL query language offered to administrate and query our Directory Service. The figure 3 describes the graphical interface offered to administrator to browse and query the directory service. It particularly shows the possibility to obtain with a simple click all the attributes of the selected entry.

Fig. 3. *Browsing Interface for querying and for administrating the directory.*

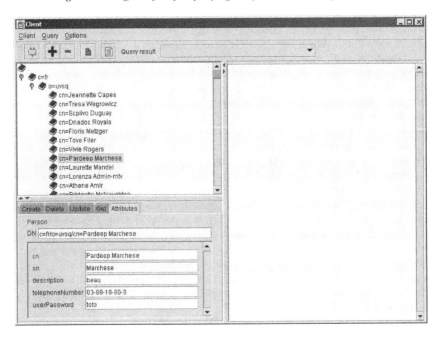

Fig. 4. *Querying facilities with DQL.*

The figure 4 illustrates the querying of information.

It retrieves the attributes of the entry named

cn=Pardeep Marchese,o=uvsq,c=fr

Since the name of this entry is not easy to write. The interface provides it to the user with a simple click on the DIT.

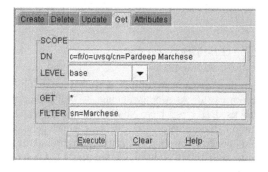

The figure 5 gives the result of the query in the LDIFF and the XML/DSML format.

Fig. 5. *Query result.*

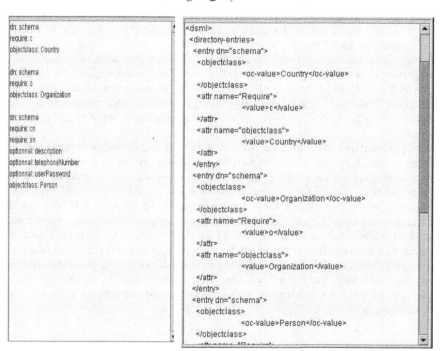

In the prototype schema information can be queried in the same way using DQL. Here we follow the classical approach used in databases or LDAP, which consists of describing the data model in the model itself what is called a meta-schema. The schema can be queried in a much more powerful way that usually provided by commercial LDAP servers. The schema can be exported as a XML/DSML document.

5 Conclusions

After the "distributed object wave", the "IETF wave" is now overwhelming the telecom world. As a consequence, LDAP protocol and LDAP servers are more and more used. As we have shown in this paper, LDAP becomes the most "lowest common denominator" to different heterogeneous technologies and services that could cooperate by sharing information through its protocol.

Due to their lightweight characteristics LDAP servers have some lacks and classical DBMS technologies such as relational and object DBMS still remain essential. Hence, this introduces new interoperability issues as data are by nature common to different applications. Several approaches have been proposed in this paper to deal with these problems, from a centralised approach with multiple views, to loosely coupled environments. Moreover, the tendency in telecom management is to promote LDAP as a management protocol of networks nodes that will be very often implemented using distributed object technology such as Corba. In this paper, we mention the need of a directory service for Corba, this answers to the connectivity problem of LDAP servers and distributed objects.

Starting from the needs of web platforms and network platforms, we analyse the requirements posed by a better integration of these technologies. The studies presented in this paper are a preliminary work in which major issues have been identified. Some solutions have already emerged and are under prototyping. We expect, during the next years, to realise as much as possible a complete and modular prototype.

Another important issue, still to be addressed, is performance which requires a good tuning of our Directory Service, as well as a good strategy for view materialization. Since it does not exist a unique good solution for all needs, a good understanding of the bottlenecks is essential to provide efficient solutions matching the needs and constraints of the applications. From our experience in the database field, this task is not easy to achieve. To better understand the problems and the way the servers must be tuned, we believe that some benchmarks should be defined; as it has been done in the classical database field. From small directory to large directory, they will be many different answers that need to be sorted out in terms of the quality of service offered versus performance. Even if the problems have been identified and solutions have emerged, we still believe that more works have to be done in order to solve them in all their generality.

Acknowledgements

The authors wish to thank Gilles Nicaud and Rémi Kerboul for their contributions on studies about replication issues in loosely coupled servers, Philippe Pucheral for comments, as well as Perrine Drumez and Agnès De La Chapelle for the developments they have realized and their great availability.

The Eurescom P925 project "Internet Middleware" must also be thank for providing very useful new LDAP requirements for web portals built with off the shelves services and for providing the opportunity to get a good skill with LDAP.

References

[1] R. Arlein, J. Freire, N. Gehani, D. Lieuwen, J. Ordille "Making LDAP Active with the LTAP Gateway" 1frst Workshop on Databases in Telecommunications/VLDB-99 Edinburgh September 99

[2] S. Cluet, O. Kapitskaia, D. Srivastara, "Using LDAP directories caches", PODS 1999, p 273-284.

[3] P. Déchamboux, R. Basset, S. Drapeau, L. Garcia, A. Lefebvre, "JORM: a Java Object Repository Mapping System", internal technical report of the RNRT Corsica project, April, 2000.

[4] T. Delot, B. Finance, Y. Lepetit, A. Ridaoui *"TINA Service Platform Management Facilities"* Networks and Services Management conference GRES99, Montréal, June 1999.

[5] T. Delot, B. Finance, "A Corba Directory Service", Electronic Journal on Network and Distributed Processing (EJNDP) n°11, March 2001 and NOTERE congress - Paris, France, 21-24 novembre 2000.

[6] T. Delot, B. Finance, "Managing Corba objects with dynamic behaviour in a Directory", Int. Symposium on Distributed Objects and Applications (DOA) 2001.

[7] DEN: Directory Enable Network www.dmtf.org/spec/denh.html 2001

[8] DMTF: Distributed Management Task Force www.dmtf.org 2001

[9] DSML: Directory Services Markup Language www.dsml.org 2001

[10] J. Fessy, B. Finance, Y. Lepetit, P. Pucheral, *"Data Management Framework & Telecom Query Service for TINA"*, fifth International Conference on Telecommunication Systems Modelling and analysis, Nashville, March 1997

[11] B. Finance, T. Delot, "CMIS-L : A Query Language for Telecommunication Management Systems", In the proceedings of 15èmes Journées Bases de Données Avancées (BDA), Bordeaux, France, p.119-137, 1999.

[12] J-R. Gruser, L.Rachid, M.E. Vidal, L. Bright, "Wrapper Generation for Web Accessible Data Sources", *Int. Conf. on Cooperative Information Systems (CoopIS)*, 1998.

[13] IETF Network Working Group RFC 2768 "Network Policy and Services: A report of a workshop on Middleware" February 2000

[14] IETF LDAP Replication Architecture Internet-draft March 2000

[15] ISO/IEC n°9594 and CCITT/X501, "The Directory models", 1990.

[16] ISO/IEC 9595 and CCITT/X710, "Common Management Information Service: CMIS", 1992.

[17] ISO/IEC/SC21/7689, "RDA: Remote Database Access", 1993

[18] H. Jagadish, L. Lakshmanan, T. Milo, D. Srivastava, D. Vista "Querying network directories" Proc. SIGMOD International Conference on Management of Data 99

[19] H. Jagadish, M. Jones, D. Srivastava, D. Vista "Flexible list management in a directory" Proc. Conference on Information Knowledge Management 98

[20] IETF M. Wahl, T. Howes, S. Kille Lightweight Directory Access Protocol (v3) RFC 2251 December 97

[21] Y. Lepetit "Overview of Data Management Issues and Experiments in TINA networks" 1frst Workshop on Databases in Telecommunications/VLDB-99 Edinburgh September 99

[22] G. Nicaud, r. Kerboul, y. Gicquel, c. Fiégel " a non-2pl locking method for coherent read without wait" 14th inforsid congress Bordeaux 96

[23] Object Management Group, "The Common Object Request Broker : Architecture and Specification", report n° 9305089, available at http://www.omg.org/, 1993.

[24] Object Management Group, "Query Service Specification", 93-3-31 1995

[25] Object Management Group, "Naming Service Specification", available at http://www.omg.org, 1995.

[26] Object Management Group, "Trading Object Service Specification", available at http://www.omg.org, 1997.

[27] Eurescom Project P925 : "Internet Middleware (for Customised Service Bundling)" at www.eurescom.de

[28] P. Pucheral, M. Abdallah "A low cost Non-Blocking Atomic Commitment Protocol for asynchronous systems 11th Conference on Parallel and Distributed Computing Systems (PDCS) Boston 99

[29] A. Sahuguet, F. Azavant, "Building Intelligent Web Applications Using Lightweight Wrappers", to appear in *Data and Knowledge Engineering*.

[30] Xpath , http://www.w3.org/TR/xpath20req .

[31] "XQuery: A Query Language for XML", http://www.w3.org/TR/xquery .

[32] T. Delot, B. Finance : "Génération de wrappers LDAP pour sources de Données relationnelles", Journées Bases de Données Avancées (BDA) 2001.

Network Convergence Using Universal Numbers:
The UPT Project

Munir Cochinwala, Harald Hauser, Naveen Suri
munir@research.telcordia.com, hhauser@telcordia.com, nsuri@telcordia.com
Telcordia Technologies
Morristown, New Jersey, USA

Abstract. One of the key technical challenges raised by the ever-closer integration between circuit-switched and packet-switched networks is address resolution of cross-network calls. In this document we describe a Pan-European project that attempts to solve the addressing problem in convergent networks using a universal, dialable number. An individual can use the number for roaming across network devices, and geographical locations.

1 Introduction

With the rapid pace of change taking place in communications technology today, business decision-makers face both compelling opportunities and potentially costly pitfalls. "Convergence" – meaning that voice, fax, data and multimedia traffic are transmitted over a single multipurpose network – is a particularly delicate issue. The business and technological advantages of combining a company's various types of communication over a common infrastructure are appealing. These advantages include:

- Decreasing transmission charges,
- Reduced long-term network ownership costs, and
- The ability to deploy a wide range of powerful voice-enabled applications.

Major data networking companies are adding IP voice functionality to their existing data products (for example, to routers and remote access servers). Conversely, voice-networking companies have expressed interest in adding data capability to their offerings. Initially, each camp will add this functionality.

One of the key technical challenges raised by the ever-closer integration between circuit-switched and packet-switched networks is the address resolution of cross-network calls. An integrated global subscriber access plan is needed. For example, the same ITU-T E.164 telephone number would reach a subscriber regardless of whether IP-based or PSTN network technologies are used. The concept of being technology independent" suggests that any global numbering/addressing plan should be abstracted as much as possible from underlying lower layer technologies.

W. Jonker (Ed.): Databases in Telecommunications II, LNCS 2209, pp. 155-166, 2001.
© Springer-Verlag Berlin Heidelberg 2001

Currently, it is possible to originate calls from IP address-based networks to other networks, but it is rare to terminate calls from other networks to IP address-based networks primarily due to the addressing. Rather, calls are generally terminated on the PSTN, so the called party can only use a terminal device connected to these networks.

In order to access a subscriber on an IP address-based network, a global numbering/addressing scheme across both PSTN and IP address-based networks needs to be developed and implemented.

ITU-T Study Group 2 (SG2) [3] is currently still studying a number of possible options whereby users in IP address-based networks can be accessed from/to PSTN users. As one of these options, SG 2 has temporarily reserved a part of the E.164 numbering resource +878 for the IP-based implementation of Universal Personal Telecommunication (UPT) services. UPT allocates a universal number to an individual. Providers of telecommunication services provide infrastructure and services to route a call to a person based on the UPT irrespective of the underlying network or device.

The simple idea behind UPT is that a user could be reached
- Anywhere (on any network)
- On any device
- On one universal phone number (UPT)

2 UPT

ETSI (European Telecommunications Standards Institute) is a non-profit organization whose mission is to produce the telecommunications standards that will be used for decades to come throughout Europe and beyond. Telephony and Internet Protocols Harmonization Over Networks (TIPHON) is a project under ETSI with the scope to standardize the interworking between traditional PSTN/ISDN/GSM networks with IP-based networks. TIPHON is developing a set of standards to support the provision of high quality telephony related services by operators such as the existing public network operators that are concerned with quality, security and call related billing. The UPT service is based on TIPHON standards. Multiple European providers are participating in the UPT trial [1]

The main purpose of the UPT project is to establish a generic, internationally interoperable IP-Telephony service based on TIPHON (Telecommunications Internet Protocol Harmonization over Networks) specifications providing functionality for:
- National and international number portability
- Personal mobility

Universal personal telecommunication (UPT) introduces the concept of personal mobility across many networks. UPT enables access to telecommunication services while allowing personal mobility [3]. It enables each UPT user to participate in a user-defined set of subscribed services and to initiate and receive calls on the basis of a personal, network-transparent UPT number across multiple networks on any fixed, mobile *or IP based terminals*, irrespective of geographical location, limited only by terminal and network capabilities. UPT allows roaming across multiple networks and devices with a single dialing plan. Mobile systems like GSM only allow roaming within their network, are device dependent and the dialing plan varies with movement.

A very important aspect for the provisioning of personal mobility is that the numbers are assigned to individuals and not to terminals and are therefore personal numbers. Each user, who wants to have his international UPT-number, has to enrol for it by an Issuing Authority (IA) only once and can keep his number for the rest of his life if he wants to. A user of this service may have one or more international number for different kinds of services. The most important requirement for the provision of number portability is a globally unique number for each user of the service, the users do not have to change even if they change their service providers (SPs) either nationally or internationally.

The proposed solution is to use UPT numbers with the following structure:

878	10	Subscriber number (10 digits)

Structure of the UPT numbers

The +878 number can be called from any Switched or Mobile Phone as well as from IP based terminals (PC and Ether phone and VoIP Terminal Adapter)

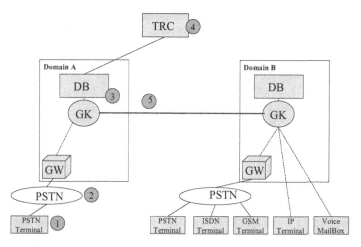

Figure 1.

2.1 UPT Architecture

The UPT architecture [5] allows each provider/domain to define their own network. Each provider manages a PSTN and IP network. Communication between the networks is implemented using gateways and gatekeepers. Currently, the protocol for communication is H.323 although SIP implementations are also under consideration. Each user (UPT numbers) has a home provider. User authorization and end-user billing are the responsibility of the provider.

The UPT architecture has a hierarchy of databases. The top-level, global database is TIPHON Resolution Capability (TRC). Each provider has a local database where the user profile including presence (current device and network) is maintained. For the purpose of this document, we assume that user authorization and presence are maintained by the home provider and within the local database. Adding multiple providers would imply that the presence information is maintained in TRC or another cross-provider database is added to the hierarchy.

TRC serves as a centralized number resolution clearinghouse to support all service providers for inter-domain call routing. Its main function is to map a UPT Number to the home service provider of that number. Inter-domain call routing follows the hop-by-hop paradigm using the two-level resolution hierarchy. The steps are as follows:

1) A call is dialed using a UPT number. Figure 1 shows the call originating from a PSTN device but it could equivalently originate from an IP device or a wireless device.
2) The local PSTN recognizes the number to a UPT number and signals to the gateway for resolution.
3) The gateway signals to the gatekeeper in its domain that queries the local db for resolution. The local db will check if this domain is the home provider for the dialed UPT number.
4) In this case, the UPT number belongs to a different domain. The TRC is queried for resolution. TRC returns the home provider of the UPT.
5) The call is established via the gatekeepers.

Note: Domain B in Figure 1 has to be aware of the current device for the dialed UPT number. The current device could be based on user profile or presence.

2.2 TRC Architecture

The TRC is functionally divided into two parts: the TRC Real-time System (TRCRS) and the TRC Administrative System (TRCAS) as shown in Figure 2. The real-time part is responsible for identifying the home service provider of the called user during the call setup process, while the administrative part is used to store and update the information about the association between subscribed UPT Numbers and the service providers. The TRC system integrates service management, number administration, number pooling, number portability and real-time translation into a single system [5].

2.3   TRC Processing Flows

Figure 2 shows functional components of TRC and processing flows from the context of subscribers and service providers. A subscriber can obtain a UPT Number from a service provider. However, the detailed number issuing process, though very important in its own right, is out of the scope of this document. Hence, those interfaces are marked as dotted lines. The GUI consists of the A1 interface with number allocation authority and the A2 interface with service providers' system. The A4 interface is for a real-time interface for providers during call set-up. The internal downloading provisioning interface, A3 between the Administrative Database (ARDB) and the Real-time Database (RTDBs) is very important, but it is internal to the TRC. It is unclear whether this interface is also a subject of standardization. This is a subject for further study.

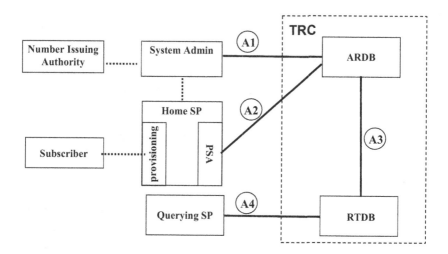

Figure 2.

The administrative part consists of a (logically) single administrative reference database (ARDB). This ARDB contains the participating service provider profiles and the mapping for each active UPT Number to its home service provider. A service provider performs number administration using its service provider service administration (PSA) system, through the TRC GUI, to create, update and de-activate records in the ARDB for the subscriber's UPT Numbers within its domain. Each service provider should be only able to access and update UPT Number mapping information for its own subscribers in the ARDB. The TRC System Administrator, acting as the number issuing authority (NAA) may also interact with the ARDB through the external provisioning database for the assignment of the UPT Numbering range/type and the validation of UPT Numbers before they are entered into the ARDB.

All mapping records are created and stored in the ARDB. Subsequent update and/or deletion of these records will also be done through the ARDB. However, during call setup, a service provider's local database and system normally does not query the ARDB directly. Instead, it queries a TRC real-time database (RTDB), which contains a copy of the mapping records of the ARDB, through the real-time query interface of TRC.

2.4 Functionality & Requirements for TRC

The detailed TRCAS functions are defined in the following paragraphs. A summary of the TRCAS functions include:

- Applying for an E.164-compliant Number by a service provider on behalf of a subscriber;
- Providing Interfaces to perform the Number Issuing Authority (by the System Administrator) functions to restrict/allocate the usage of UPT numbers;
- Providing for the assignment and withdrawal of UPT numbers;
- Providing a database for the registration of UPT numbers;
- Managing the resolution for a E.164 compliant Numbering Plan;
- Providing interface and implementation for Number Portability;
- Facilitating Download Interface to download information to the RTDB.

An UPT subscriber who wants to have numbers assigned must order the service from a provider of his choice. Once the customer places an order to a service provider for the UPT service, the service provider registers the subscriber into the TRCAS. The service provider uses TRCAS' GUI to choose a subscriber's UPT. The registering service provider, called the Home Service Provider, is responsible for the management of the records. The functions for the management and administration include data entry, and record changes. The resolution information will be effective within 12 hours after the completion of a successful provisioning transaction.

2.4.1 Number Assignment & States

The following rules apply for the assignment of a UPT number:

- The total available numbers are from 878010000000000 to 878019999999999.
- A UPT number will not be assigned to more than one subscriber.
- A customer/subscriber can request more than one UPT number.
- A customer/subscriber may be allowed to request a specific number.

- If a service provider requests a specific number, the TRC ARDB will check the availability of the number and assign it to the provider, if available
- The TRC ARDB can choose an available number if the customer/subscriber has no preference.
- The TRC ARDB will be able to reserve multiple numbers by a service provider. The number generated by the system may or may not be consecutive.

A UPT Number in the TRC ARDB database will have one of the following statuses:

- AVAILABLE: This status is assigned initially for all available numbers to be used for the UPT numbering plan or new additional numbers to be added to the pool of available numbers. It is a state that indicates the number is not used, not restricted for use, and not allocated to any service provider or subscriber.
- RESERVED: This status indicates the number is reserved by a service provider for a potential subscriber but no active subscriber for the number exists; and therefore, the number is not yet in the RTDB.
- ACTIVE: This status indicates the number is assigned to a subscriber. When the provisioning of a number is completed, there will be resolution information associated with the number in the TRC ARDB and in the RTDB.
- UNAVAILABLE: This status is assigned and released manually by a system administrator. This status will only be used when a number needs an administrator's manual attention and to make it unavailable for regulatory reasons.

A number starts out in available state. When a service provider first reserves the number (1) it moves into reserved state. The service provider can assign the number to a subscriber, complete necessary provisioning, and the status moves to an active state (2). During this transition, information is sent to the RTDB. As more updates are associated with the number (3) the number stays in an active state, and the changes are updated according to the download interface to the RTDB. For this implementation stage, no number can be unreserved, therefore, the number status will not return to an Available state from a Reserved state.

2.4.2 Number Portability

The number portability operation [9] is required when a user moves from one service provider to another. The user maintains his UPT. The UPT number portability is much simpler than traditional number portability. In the UPT system, the old and new providers have to agree and the internal TRC databases (real-time and administrative) have to be updated to reflect the change. In traditional number portability, databases of all providers need to be updated so that valid call routing can be implemented.

Number porting sequence is as follows:

- In response to a request for a UPT number port from either a subscriber or potential recipient, the donor service provider can use the UPT Update screen to obtain a Porting Authorization Pin Number from the TRC system for one or more particular UPT(s). The Home Service Provider of a UPT can only request a PIN number.
- The donor service provider will give the PIN Number of the UPT to be ported to the potential recipient service provider via email, phone or FAX.
- The recipient service provider will use the UPT Update GUI to provision the ported UPT number. The PIN number given by the donor service provider supplies the authorization for the transfer of the ownership of the UPT from the donor to the recipient.
- Once the recipient service provider completes the porting and related provisioning, the recipient will be reminded to inform the donor of the completion of the transfer via email/phone/FAX.

2.4.2 Performance & Availability Requirements

The TRC system has stringent requirements for performance and availability. The requirements are specified to ensure worldwide carrier-grade operations. We outline a few of the requirements.

2.4.2.1 Availability

The availability of the real-time part of TRC depends on the availability of an RTDB, the number of RTDBs in the network, and the accessibility of those RTDBs to service providers.

The TRC real-time system has the following requirements for availability:

- The system will have 99.999% availability.
- The scheduled service time of the real-time part of TRC will be 24 hours per day, 7 days per week i.e., there is no scheduled downtime for the real-time part of TRC.
- The average service downtime of an RTDB should be no more than 12 hours, cumulative, per year.

The TRC administrative system has the following availability requirements:

- There will be a web interface and a database server with processing capability of 99.5% availability.
- The scheduled service time of the administrative part of TRC shall be 24 hours per day, 7 days per week with scheduled maintenance down time no more than 5 hours per week.

2.4.2.2 Capacity

The TRC system shall provide the initial capacity to handle 10 million numbers with the expandability to a potential target of 10**9 =10 billion numbers.

2.4.2.3 Performance

The real-time system shall have the capability to handle 30,000 queries per second.

The TRC administrative system shall have the capability to handle 1000 transactions per day for the UPT registration and provisioning activity. Estimated peak hour load is 30% of one day transaction = 300 transactions. Estimated peak minute load is 20% of the peak hour load = 60 transactions.

The TRC administrative system shall have the capability to handle 100 concurrent users.

3 Discussion

The UPT system attempts to reduce complexity of the next generation world by providing a unified addressing scheme. This scheme allows users the flexibility to roam across multiple networks, introduce new devices into their environment and be able to receive/send communication without getting a new address. Furthermore, it reduces an individual's dependency on service providers by providing worldwide mobility. However, the UPT system introduces new problems that have to be solved before it can claim success. None of the problems seem insurmountable. We outline a few of the problems:

3.1 Network & Database Traffic

Currently, only logical numbers such as 800 and mobile numbers require translation to current location and device. These numbers comprise of only 25% of the total network traffic. [10] If UPT is fully deployed; all calls will have to be translated. This will increase database and network traffic significantly. The size of the network database and database queries increases by a few orders of magnitude. [10]

3.2 End User Billing

UPT allows worldwide roaming across networks. The calling user needs to know how much a call is going to cost due to worldwide roaming. Multiple options such as the GSM model are being considered.

3.3 Settlement

Providers will be sharing resources and routing calls for roaming UPT users that do not belong in their domain. The OSP protocol is the mechanism for settlement across providers both on a call-by-call basis and by agreements. [7]

3.4 Privacy & Security

Privacy of user data and security of the system (especially TRC) are a big concern. A global database has to be secure from denial of service attacks and yet be accessible worldwide for updates from legitimate users.

3.5 Quality of Service

The network to ensure PSTN like quality of service has to be implemented. Admission control can be implemented to handle maximum number of phone calls. However, future services such as streaming video may require more complex policies.

From a business perspective, the service providers will realize a major cost savings using the IP network to provide a myriad of services. That will impact the profitability, which could be passed on as savings or new services to their customers. The service providers are still faced with the challenge to calculate the net impact of traffic substitution from traditional PSTN to VOIP. The challenge is in revenue reduction with increased profitability due to the cost saving incurred from using IP networks. This is one of the major business challenges that are being debated. All of this of course assumes that IP network brings in cost savings over the traditional network.

4 Status & Future Steps

The system has completed a trail phase and is now in a pre-commercial phase. A network and initial deployment of the UPT service components have been implemented. Multiple European providers such as Telekom Austria, Telefonica, OTE and Tele Denmark are part of the pre-commercial phase.

It is expected that operators will start deploying the UPT service in the fourth quarter of 2001. The deployment plan varies by operators. Other European operators are considering deploying the service also and we are starting to work with US operators for UPT service in the US.

New services such as using a UPT for unified messaging, media translation based on device characteristics are being considered. These services need to be defined and implemented. A key aspect of UPT is provider independence. Each provider is free to deploy new services as needed. Providers can use new services on their networks for customer retention and gain. It also allows providers an entry point in the IP world and Internet telephony.

The UPT service provides great value to end-users in reducing complexity of dialing plans, independence from multiple addresses for multiple devices and networks with low cost. It has the appearance of 'big brother' since the providers will know exactly where to find a person and has privacy and security implications. The success and failure of the UPT service will be determined by customer acceptance.

5 References

[1]   TIPHON UPT Phase 1 Service Description (http://www.etsi.org/tiphon).
[2]   ITU-T E.164: "Numbering Plan for the ISDN Area"
[3]   ITU-T E.168: "Application of E.164 Numbering Plan for UPT"
[4]   ITU-T Q.1290: "Glossary of terms used in the definition of intelligent networks"
[5]   TIPHON DTR 02003: "TIPHON Architecture",(http://docbox.etsi.org/Tech-Org/TIPHON/Document/tiphon/07-drafts/wg2/ DTR02003/)
[6]   TIPHON DTR 04006: "TIPHON Call Routing on IP Telephony Networks" , (http://docbox.etsi.org/Tech-Org/TIPHON/Document/tiphon/07-drafts/wg4/ DTR04006/)

[7]   TIPHON DTR 03004: "TIPHON Open Settlement Protocol (OSP)" (TS 101
         321),   (http://docbox.etsi.org/Tech-Org/TIPHON/Document/tiphon/07-
         drafts/wg3/ DTR03004/)
[8]   Stage 1 Service Description for the non real-time part of the Global IP-
         Telephony   Database   Service   (TRC),   (http://docbox.etsi.org/Tech-
         Org/TIPHON/Document/tiphon/07-drafts/wg2/ DTS02003/)
[9]   TIPHON DTR 4007 "Number Portability and its Implications for TIPHON",
         (http://docbox.etsi.org/Tech-Org/TIPHON/Document/tiphon/07-drafts/wg4/
         DTR04007/)
[10] ICDE 2001, "Database Performance for Next Generation Telecommunications,"
         Munir Cochinwala

Toward Universal Information Models
in Enterprise Management

Jean-Philippe Martin-Flatin

AT&T Labs Research, 180 Park Avenue, Bldg. 103, Rm. A001,
Florham Park, NJ 07932, USA
jp.martin-flatin@ieee.org

Abstract. The DMTF's recent work on management information modeling in the IP world has highlighted that a number of problems are still unsolved in this important area of enterprise management. In this paper, we identify five: finding the right level of abstraction, building on past experience, devising good models, finding a good trade-off between quality and timeliness of new models, and attracting the best experts in the field in standardization efforts. We propose to alleviate them by splitting information modeling into two phases that involve different people with different skills. In the first phase, designers and experts in a given technology (be it a router, a service, a policy, etc.) capture the core issues for managing it in a Universal Information Model (UIM) that is independent of any management architecture. At this stage, low-level engineering details are ignored. In the second phase, code-oriented engineers instantiate the UIM into a data model specific to a management architecture (e.g., an SNMP MIB or a CIM schema). These people are specialists of SNMP or WBEM, but are not necessarily experts in the technology being managed.

Keywords: Information Model, Data Model, Internet, Management.

1 Introduction

In the Internet Protocol (IP) world, since the early 1990s, network management has been dominated by the Internet Engineering Task Force (IETF)'s management architecture, named after its communication protocol: the Simple Network Management Protocol (SNMP) [12]. In the meantime, systems and application management have mostly relied on proprietary solutions. Service and policy-based management are still in their infancy: standards are still being defined, and are thus not yet widely supported by deployed equipment. To date, integrated enterprise management is still wishful thinking.

This situation may change in the near future. The Distributed Management Task Force (DMTF) is currently working on a new management architecture: Web-Based Enterprise Management (WBEM). This alternative to SNMP management encompasses the entire realm of enterprise management: network element management, network management, systems management, service management, application management, policy-based management, etc. The main strengths of WBEM over SNMP

W. Jonker (Ed.): Databases in Telecommunications II, LNCS 2209, pp. 167-178, 2001.

are its object-oriented information model—the Common Information Model (CIM) [2]—, the large scope of management areas that it attempts to model (as opposed to SNMP's focus on network element management), and its interest in low-level, machine-oriented management abstractions as well as high-level, people-oriented abstractions—unlike SNMP, which is characterized by instrumentation Management Information Bases (MIBs). These qualities, plus others that are not all of technical nature (e.g., the fact that it is backed by most of the major vendors in the IT and networking industries), make WBEM a serious contender of SNMP for this decade.

So far, most of the DMTF's work has focused on information modeling—namely, the definition of the CIM Core and Common Models [2]. This sudden rash of activity in an area that had remained fairly quiet for several years has unveiled a number of problems for management-application designers and domain-specific modelers. We identified five. Interestingly enough, none of them are specific to WBEM or CIM.

First, finding the right level of abstraction for an information model is not an easy task. It is quite difficult to devise a model that is neither cluttered with low-level engineering details, nor overly generic and abstract.

Second, over the years, the open management community has not built a reputation of being immune to the *reinvent the wheel* antipattern—a pervasive disease in software engineering. In the recent past, the DMTF has produced intense efforts in information modeling; but some of its work was a waste of time, for it was duplicating previous efforts by the IETF in the same field. Similarly, throughout the 1990s, the two main standardization efforts in management—Open Systems Interconnection (OSI) management and SNMP—have, to a large extent, followed parallel tracks with few constructive interactions.

Third, a number of information models are not good enough. Some mistakes are minor, but others may be time bombs. For deployment reasons, correcting a bogus or incomplete model takes a lot of time—so much so that in practice, the market is generally stuck for years with whatever information models have been standardized, be they good or bad. Standards bodies should therefore pay better attention to devise good models in the first place. The main causes of this problem are delineated in the next two problems.

Fourth, in the design of management information models, "fast is not beautiful", so to say. Some information models leave a lot to be desired because they were moved too quickly through the standardization process, in order for vendors to swiftly put their software products on the market.

Fifth, researchers, especially from academia, have had little impact on the standardization of management information models in the IP world. Instead, the IETF and DMTF models have traditionally been devised by the engineers in charge of designing and coding management applications. (There are only few exceptions to this rule.) More generally, standards bodies have often proved unable to attract the best technology experts in the field. This has contributed to the flaws and limitations exhibited by some information models.

In this article, we propose to alleviate these five problems by splitting information-modeling efforts into two phases, which involve different people with different skills and different goals. In the first phase, information modelers, versed into abstraction and architectural aspects, produce a Universal Information Model (UIM) that is independent of any management architecture. These experts in a specific technology—e.g., in IP routers, differentiated services, Quality-of-Service (QoS) provisioning, Service-Level Agreements (SLAs), or IP telephony—capture the core issues for managing this technology, and ignore low-level engineering details. In the second phase, engineers versed into programming instantiate the UIM into a data model that is specific to a management architecture—e.g., an SNMP MIB specified in SMI (Structure of Management Information [8]), or a CIM schema expressed in MOF (Managed Object Format [3]). These people are specialists of a management architecture (e.g., SNMP or WBEM), but are not necessarily experts in the technology being managed.

We believe that two-tier information modeling should not be specific to the IP world, and would equally improve the quality of management information models outside the IETF and DMTF realms (e.g., for fixed and cellular telephone networks relying on OSI management). But for the sake of clarity, the scope of this article is limited to enterprise management in the IP world.

The remainder of this paper is organized as follows. First, we detail the problem statement. Next, we shed some light on our vision of two-tier information modeling. We then present a research agenda for UIMs and investigate related work. Finally, we summarize this article and present directions for future work.

2 Five Problems in Management Information Modeling

Let us now dive into the problems outlined in the introduction and show their interdependencies. Readers who are not familiar with the way management information is currently modeled by the IETF and DMTF are referred to [6]. In particular, this technical report summarizes the languages used to express data models (MOF for WBEM, SMI for SNMP, and SPPI for policies), the communication protocols used to transfer messages (SNMP, HTTP for WBEM, and COPS for policies), and the languages used for representing and encoding management data in these messages (BER for SNMP and XML for WBEM).

Problem 1: Between overly generic and cluttered models

Finding the right level of abstraction for management information models is a difficult exercise. There are many options. At one end of the spectrum, some theoreticians produce overly complex models. By trying to be too generic, they define extremely abstract concepts that are not very useful to the market. An example of this is the Object Management Group (OMG)'s four-tier meta-model architecture [9]: It started as a much needed solution to a real problem, but it is so complex that few people really understand how to use it efficiently in practice.

At the other end of the spectrum, we find developers that are only concerned with programming. They clutter information models with so many details that the big picture of the models is blurred, and sometimes even hidden. This approach is epitomized by SNMP MIBs: their design often looks cryptic to information modelers. We claim that one cannot understand the convoluted design of some MIBs without a full understanding of the limitations of SMIv2, the language used to specify them. For instance, the definition of multi-dimensional tables in a MIB is severely constrained by the notion of *conceptual tables* in SMIv2 [11] and the lack of support for nested tables in this language. To give another example, SMIv2 does not support the remote invocation of actions (*à la* OSI) or methods (*à la* Java RMI), but relies instead on an ugly kind of programming by side effect, whereby setting an integer to different values allows a manager to trigger different actions on an agent [5]. In short, information models in the SNMP world are often weirdly structured—at least to the eyes of non-SNMP information modelers—because of the limitations in the language used to express them.

Note that cluttered models are not specific to SNMP. In WBEM, the language used to express information models (MOF) is considerably richer than SMIv2, but this does not prevent CIM models from being cluttered with low-level programming details. An example of this was a recent discussion of the DMTF Events WG on whether a new property (i.e., an attribute in standard object-oriented parlance) called `Addition-alText` should be added to CIM events, as some organizations felt that the `Description` property, currently used to describe events, was insufficient in some cases [10]. Such details are very relevant to application developers, and completely irrelevant to information modelers. Another example is the Unified Modeling Language (UML [9]) diagrams that the DMTF uses to depict its information models. Although these diagrams can be useful to get the big picture of a model, they are often very detailed and fairly difficult to understand for a person with no or little knowledge of CIM. In particular, experienced information modelers with an SNMP or OSI background sometimes find it difficult to map their knowledge onto the DMTF's information models, especially when a well-established terminology is changed (see Problem 2).

Problem 2: The *reinvent the wheel* antipattern

In 1998, the DMTF moved from desktop management to enterprise management. Since then, its information-modeling activities have been thriving. By and large, this work has been, and is still, performed independent of the IETF's. Few people belong to both DMTF and IETF WGs. As a result, cross-pollination between these two communities is rare, and the know-how accumulated by the IETF in the 1990s is often ignored by the DMTF[1]. We believe that this is a typical occurrence of the *reinvent the wheel* antipattern [1].

---

[1] There are a few notable exceptions, e.g. in policy-based management. But the situation is even worse if we consider the lack of cross-pollination with other standards bodies that have a respected expertise in management, e.g. the TeleManagement Forum (TMF), the International Telecommunications Union—Telecommunications Sector (ITU-T), the Object Management Group (OMG), and the International Organization for Standardization (ISO).

Clearly, this is counterproductive, as the DMTF would be better off working on new issues, or issues ignored so far by the IETF. Worse, it also enables old problems to resurface, and requires new engineers to go through the same long and painful learning process that their SNMP elders went through a few years ago. A third problem is that the terminology keeps changing, as each new WG starts from scratch and redefines well-established concepts (e.g., see the new meaning given to the terms *event* and *notification*, and the intrusion of the concept of *indication* in the CIM 2.5 event model [4]). These terminological changes often cause confusion, especially when SNMP- and CIM-based management systems co-exist in the same Network Operations Center (NOC).

Problem 3: Some models are not good enough

The third problem is that some information models leave a lot to be desired. An example of this is the Address Translation Group, defined in MIB-I (RFC 1156) and deprecated shortly afterward in MIB-II (RFC 1213), because a major flaw had been discovered during the deployment phase (see RFC 1213, Section 3.6). Another example is the lack of per-interface access control lists in MIB-II, which has led many vendors to define their own in proprietary MIBs—or even worse, to demand that monitoring staff use the Command Line Interface (CLI) via `telnet`. A third example is the CIM event model, currently under revision, which promotes a confused idea of the severity of an event. The severity normally indicates whether an event is perceived to be critical, serious, or simply informative by the sending party. It gives a hint to the receiving party whether this event should be processed urgently. Indeed, the DMTF defines the severity levels `Information`, `Warning`, `Minor`, `Major`, `Critical`, and `Fatal`. But the WG recently accepted that a new severity level called `Clear` should enable a sending party to inform the receiving party that a certain problem has disappeared [10], which defeats the purpose of an event's severity.

This list of examples is by no means exhaustive, and a number of other SNMP MIBs and CIM schemas have well-known flaws or limitations. The two main causes that we identified for Problem 3 are detailed in Problems 4 and 5.

Problem 4: Fast design vs. good design

In our view, the main source of shortcomings in management information models is that the IETF and DMTF standardize these models too quickly. One reason for this is that the WGs who define information models are mostly driven by vendors, and in this business, vendors traditionally value speed over quality. Another reason is an overreaction to the slow pace of OSI management standardization in the early days of open management.

Vendor-driven WGs value speed over quality: In the IT and networking industries, vendors operate in very competitive markets, and they work hard to be the first to support a new technology. Being first is good for their image, their profits, and their stocks—sometimes irrespective of the quality of their implementation. Because a number of customers do not buy a technology unless it can be managed, most vendors release a new technology only once its management software is ready. Needless to say, management information modelers must meet drastic time schedules, squeezed between the time the technology is ready and the time it can be deployed. To meet these deadlines, they spend less time designing an information model than implementing it: they jump too quickly to engineering details. To make things worse, they also have an intense sense of competition with the other companies present in the WGs in charge of standardization. These are two strong incentives for designing information models very fast, and trying to rush them through the standardization process demanded by some of their customers.

The life span of a management information model is usually much larger than the time it takes a WG to design it; a typical ratio is 1:10. Consequently, saving one month on the design of important building blocks of an information model makes no sense in the long term. This small saving can make modelers overlook important aspects of the model and require, several years later, changes in the model and re-deployment of management software in the installed based—a major endeavor that can be immensely expensive for vendors and customers alike.

Overreaction to OSI management: In the late 1980s and early 1990s, the ISO and ITU-T standardized very slowly in the management field: standardization cycles took four years. To a certain degree, the way management information is currently modeled by the IETF is a reaction to the slow pace of OSI management information modeling in those days. The management community has been durably marked by this phenomenon, and the IETF and DMTF WGs are still afraid of engaging in never-ending discussions, seeking an elusive consensus. So they keep all design-related discussions short—even when the management issues are tough and demand long discussions. We refer to this remanence effect as *overreaction to OSI management*.

The "religious wars" between OSI management and SNMP are long past. WBEM has nothing to do with these wars. It would be useful for the management community to rid of this conditioned reflex, which might have contributed in the past to the success of SNMP-based management, but now causes more problems than it solves.

Problem 5:   The best technology experts are rarely involved in standardization

Another important cause of Problem 3 is that the best technology experts in the field rarely participate in the WGs in charge of management standardization (and this goes beyond the sole IETF and DMTF). We see two reasons for that.

The big picture is blurred: Experts generally stay away from standardization efforts in management information modeling because low-level details are of no interest to them. They are more interested in defining a sound backbone and getting the main classes and relationships right. To an information modeler, it can be frustrating to have his/her model severely limited by a modest information meta-model, or constrained by the language used to express the information model—this language being supposedly "demanded by the market". In this respect, SMIv2 can be an efficient repellant for information modelers that are not versed into SNMP management application programming. The situation can even become intolerable for academics when politics get in their way—and vendor-driven WGs are not always immune to politics...

Fast vs. smart design: Most researchers prefer to do things right than fast. This objective is incompatible with the fact that WGs are mostly driven by vendors, who favor speed over quality.

To conclude with the problems in information modeling, we emphasize that these five problems are *not* specific to the DMTF or IETF. For instance, some of them were not unheard of at the ITU-T, a decade ago, when the management and control of fixed telephone networks were undergoing standardization.

3 From One- To Two-Tier Information Models

In our view, these five problems share two root causes:

– With one-tier information models, we try to do too many things in one step, and require too many skills from the same people. Instead, we propose to adopt two-tier information models. These tiers are devised by different people with different skills.
– The management issues for a given technology are intrinsically independent of the architecture (SNMP or WBEM) used by a customer to manage this technology. Technology-specific information models should therefore contain an architecture-independent core.

Based on these two important remarks, we propose to change the way information modeling is currently performed in the IP world.

3.1 Two tiers: one UIM, several data models

Today, the IETF and DMTF go directly from a high-level description of the managed technology to very detailed SNMP MIBs or CIM schemas. Instead, we propose to split information modeling into two phases (see Fig. 1). First, for each technology to manage, we define a Universal Information Model that is independent of SNMP and

WBEM. Second, from this UIM, different WGs derive different data models in the form of SNMP MIBs, PIBs, CIM schemas, etc.

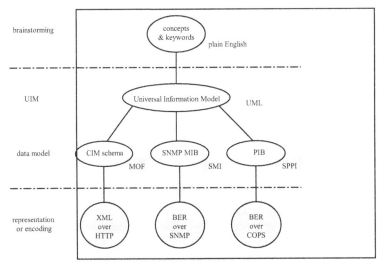

Fig. 1 Two-tier information model

Universal Information Models: Today, the management community needs good information models that management application developers, information modelers, administrators, and operators can easily understand, without having to master the sometimes cryptic notations of SMI, MOF, GDMO, etc. To achieve this, we propose to define a single Universal Information Model (UIM) for each technology to manage. This makes up the first tier of our two-tier information model (see Fig. 1).

A UIM is an object-oriented abstract model for managing a technology. It consists of object-oriented classes and relationships organized in UML diagrams. The management concepts modeled by these diagrams are identified by WGs including some of the best worldwide experts in this technology.

UIMs convey the big picture of the models to people; they ignore the details. They are not meant for machines or compilers, but for people who are familiar with the technologies being managed, e.g. information modelers and administrators. This level of abstraction is independent of the idiosyncrasies of any specific management architecture. In particular, UIMs are not constrained by the rather limited SMIv2 language.

We propose to adopt the UML meta-model [9] for expressing UIMs because it is standard and object oriented. It is also fairly close to the DMTF meta-model.

UIM standardization should be driven by joint IETF and DMTF WGs, and may also involve people from the TMF, the OMG, the W3C, etc. These WGs should not be driven solely by vendors. Academic researchers, who are less prone to having vested interests in standardization activities than people from industry, should get involved too.

Data Models: The second tier of our information model consists of data models derived from the UIM (see Fig. 1). In SNMP, data models are usually SNMP MIBs (or PIBs in the case of policies). In WBEM, they are CIM schemas. A data model is a level of abstraction well suited to developers of management applications. It is specific to a given management architecture and contains many details that should be hidden to the end-users of these applications—especially administrators and operators in NOCs.

Data models directly reflect the constraints of the language used to express them. For instance, the expressiveness of MOF is greater than SMIv2's, so CIM schemas can be more expressive than SNMP MIBs. Another example is that data models need not necessarily be object oriented, although UIMs are. Note that our proposal does not prescribe the language that should be used for defining a data model: this is left to the management architecture.

The main actors for defining data models should be vendors developing real-life management applications (managers or agents).

3.2   Advantages of using two-tier models

Our proposal alleviates the five problems described earlier.

Between overly generic and cluttered models: By adding the UIM-design phase between the brainstorming and data-modeling phases, we make sure that information modelers are not hampered by low-level details when they devise a new model. This decreases the risks of making a bad design mistake in the information model, and of having to update it once it is already deployed.

The **reinvent the wheel** *antipattern:* We make it possible to solve Problem 2 by devising good, solid UIMs that are independent of any specific management architecture and can live for many years, without undergoing major upheavals. When a new management architecture is defined, only data models need to be defined. By doing this, we help build on past experience and help prevent old problems that have already been solved from resurfacing in the future. We also prevent customers from being confused by constant changes in the terminology.

Some models are not good enough: We alleviate this problem by addressing the next two.

Fast design vs. good design: We solve Problem 4 by having vendors compete on the fast implementation of data models rather than on the definition of UIMs. Mapping a UIM onto a data model should not take much time, especially as the DMTF and IETF WGs develop some know-how in doing so.

The best technology experts are rarely involved in standardization: We offer a solution to Problem 5 by making standardization efforts a lot more attractive to experts, especially academic researchers. With UIMs, they are no longer drown by low-level details and can focus on getting their models right reasonably fast. And by bringing more experts into standardization, we increase the chances of devising good models in the first place, and decrease the risks of having to update an already-deployed model.

4   A Research Agenda for UIMs

For UIMs, the first item on the research agenda is probably to systematically formalize existing SNMP MIBs in the form of UIMs, to abstract UIMs out of the existing DMTF UML diagrams (most notably, by making them CIM independent), and to attempt to merge these pairs of UIMs, technology by technology. The primary outcome of this work would be to make it possible for information modelers not interested in the idiosyncrasies of SNMP and CIM to study, and possibly improve, these UIMs, and to document different, incompatible approaches between different management architectures. Another outcome would be to highlight the areas where the DMTF's CIM schemas are lagging behind the IETF's SNMP MIBs, and those where the DMTF expands on existing SNMP MIBs. These two outcomes would be particularly useful to customers, as they would enable them to compare different approaches to a single problem, and to make an educated guess when selecting SNMP or CIM for deploying a new management solution.

It would also be interesting to assess whether deriving several data models from a single UIM eventually facilitates the translation between these data models. Past work has shown the difficulty to translate managed objects expressed in SMI, GDMO, and IDL, and the semantic losses that can occur [7]. Do shared UIMs alleviate these problems?

Another point worth investigating is whether UIMs need the equivalent of the CIM Core Model in order to share some building blocks. Should derived types such as Ipv4Address, Ipv6Address, Date, and TimeSeries be defined once and for all, and shared by all UIMs?

5   Related Work

Our proposal can be viewed as a generalization of a mechanism used by the DMTF WGs. The main differences are fourfold:

- These WGs generally spend much more time specifying the nitty-gritty of the data models than putting together smart UML diagrams. Some WGs only occasionally update their UML diagrams. In our approach, the definition of the UIM is a goal *per se* that usually takes more time than deriving the data models.
- Usually, at the DMTF, the same people are in charge of defining the UML diagrams and the data models, which requires a mix of skills rarely found in practice. In our approach, these two teams are different.
- The UML diagrams that are produced by the DMTF WGs are completely CIM specific. In two-tier information modeling, the UIM is independent of CIM.
- Finally, the DMTF is using UML diagrams only for improving the quality of the models. We define UIMs also for sharing them with other standards bodies such as the IETF.

6 Conclusion

We have exposed five problems associated with information modeling in the IP world, and proposed to alleviate them by splitting information modeling into two phases. In the first phase, designers and technology experts define a Universal Information Model (UIM) that is independent of any specific management architecture. A UIM focuses on the big picture of the management issues for a given technology. It is expressed in UML. In the second phase, different people derive multiple data models from a single UIM. In SNMP, these data models are MIBs written in SMI, or PIBs written in SPPI; in WBEM, they are CIM schemas written in MOF. For a given technology, a UIM and its associated data models constitute what we call a *two-tier information model*.

Some work is currently under way at AT&T Labs Research to define UIMs for policy-based management and IP-router management. In particular, we have begun reverse-engineering and merging the existing data models of the IETF and DMTF.

Acknowledgments

Several ideas presented here were debated at the 8th meeting of the Network Management Research Group (NMRG) of the Internet Research Task Force (IRTF). The clarity of this paper benefited greatly from these discussions. The author would also like to thank C. Kalmanek for his useful comments on this article.

References

1. W.J. Brown, R.C. Malveau, H.W. McCormick III, T.J. Mowbray. *Antipatterns: Refactoring Software, Architectures, and Projects in Crisis*. Wiley, New York, NY, USA, 1998.
2. W. Bumpus, J.W. Sweitzer, *et al*. Common Information Model: Implementing the Object Model for Enterprise Management. Wiley, New York, NY, USA, 2000.
3. DMTF. Common Information Model (CIM) Specification. Version 2.2. June 1999.
4. DMTF Events Working Group. *Common Information Model (CIM) Indications (Final Draft). Version 2.5 Final Draft*. DMTF, December 14, 2000. Available at: http://www.dmtf.org/var/release/Whitepapers/DSP0107.htm
5. J.P. Martin-Flatin. *Web-Based Management of IP Networks and Systems*. Wiley, Chichester, UK. To appear in 2001.
6. J.P. Martin-Flatin. *Toward Universal Information Models in Enterprise Management*. Technical report TD-4XURBN, AT&T Labs Research, February 2001.
7. S. Mazumdar. "Inter-Domain Management Between CORBA and SNMP". In Proc. 7th IFIP/IEEE International Workshop on Distributed Systems: Operations & Management (DSOM'96), L'Aquila, Italy, October 1996.
8. K. McCloghrie, D. Perkins, and J. Schoenwaelder (Eds.). *RFC 2578. Structure of Management Information Version 2 (SMIv2)*. IETF, April 1999.
9. OMG. OMG Unified Modeling Language Specification. Version 1.3. OMG, March 2000.
10. C. Shaw (Ed.). Minutes of the DMTF Events Working Group meeting of Jan 11, 2001.

11. R. Sprenkels and J.P. Martin-Flatin. "Bulk Transfers of MIB Data". *The Simple Times*, 7(1):1-7, 1999.
12. W. Stallings. *SNMP, SNMPv2, SNMPv3, and RMON 1 and 2*. 3rd edition. Addison-Wesley, Reading, MA, USA, 1999.

Author Index

Lecture Notes in Computer Science

For information about Vols. 1–2104
please contact your bookseller or Springer-Verlag